2

GENETICS – RESEARCH AND ISSUES



# RESISTIN

# STRUCTURE, FUNCTION AND ROLE IN DISEASE

# GENETICS – RESEARCH AND ISSUES

Additional books and e-books in this series can be found on Nova's website under the Series tab.

GENETICS – RESEARCH AND ISSUES

# RESISTIN

## STRUCTURE, FUNCTION AND ROLE IN DISEASE

GERALD MALDONADO

EDITOR

nova science publishers
New York

**Library of Congress Cataloging-in-Publication Data**

ISBN: 978-1-53614-543-4
Library of Congress Control Number: 2018958560

*Published by Nova Science Publishers, Inc. † New York*

# Contents

# PREFACE

Over the past decade, adipose tissue has received much attention for its immunological properties. Additionally, it has become increasingly apparent that adipose-secreted factors such as leptin, tumor necrosis factor-a, IL-6, adiponectin and more recently, resistin are implicated in some inflammatory and metabolic diseases. As such, Resistin: Structure, Function and Role in Disease opens by discussing resistin's therapeutic properties. Because evidence appears to suggest that resistin is a proinflammatory cytokine, resistin may impact metabolic disease. Following this, the author's present experimental and clinical evidence indicating that resistin may have functions related to inflammation and oxidative stress. The circulating resistin level may be also useful as a marker of the effects of exercise. Resistin may also encourage vasculopathy through the promotion of macrophage scavenger receptors and CD36 atheroma macrophages expression, the formation of foam cells, as well as proliferation and migration of vascular smooth muscle cells. This causes endothelial dysfunction, resulting in limited nitric oxide and increased superoxide production. When present in the diseased vessel wall, resistin may orchestrate pro-atherosclerotic events, contributing to atherosclerosis progression towards plaque rupture. Thus, resistin may augment monocyte and macrophage infiltration, interfere in cell cross-talk and modulate the expression of various inflammatory cytokines leading to the acceleration of plaque growth. Based on the known sequential events of obesity preceding severe acute pancreatitis, the authors investigate resistin as a new potential predictive marker of fat necrosis and pancreatitis severity.

Chapter 1 - Over the past decade, adipose tissue has received much attention not only for its endocrine properties but equally also for its immunological activity. In consideration of this, it has become increasingly apparent that adipose-secreted factors such as leptin, tumor necrosis factor (TNF)-a, IL-6, adiponectin and more recently, resistin appears implicated in some inflammatory and metabolic disease. Resistin, in particular, has received meticulous attention as a potential factor in obesity-mediated type 2 diabetes mellitus (T2DM), Metabolic Syndrome, and Cardiovascular. Serum levels of resistin, an adipose tissue-derived adipokine, are increased in human obesity and are positively correlated with atherosclerotic cardiovascular disease. As such, several notable questions remain to be answered within the field of resistin biology. For instance, what conclusion can be drawn with regards the physiological function of resistin? Do the studies over the past year evaluate a role for resistin in the pathogenesis of obesity-associated inflammation and T2DM, Metabolic Syndrome, and Cardiovascular, or merely add to the chaos and confusion in the existing literature? These chapter results reveal that resistin is a highly attractive therapeutic target in factors related to diseases humans. Because current evidence appears to suggest that resistin is a proinflammatory cytokine. Thus, like many other adipocytokines, resistin may possess a dual role in contributing to metabolic disease: first through its direct effects on substrate metabolism and second, through regulating inflammation within its target tissues. The chemistry of resistin has also been the subject of investigation and like adiponectin, the homo-oligomerization of this protein has a bearing on its function. This chapter reviewed to summarize data about structure and function resistin. Also, role and contribution of resistin in the initiation and progression of pathophysiology actions among human and other mammalians reviewed. Summarize data have obtained from clinical researches and animal studies.

Chapter 2 - Resistin was discovered as an adipocyte-driven molecule, and human resistin is predominantly secreted in the macrophages, with a positive connection with cardiometabolic pathophysiology. The biology of human resistin appears to differ from that of animal resistin, and the function of human resistin remains unknown. Using a MEDLINE/Pubmed database, the authors summarized the English literature investigating the effects of exercise intervention on the circulating resistin levels in humans. Regular exercise (regardless of modality) generally reduces the circulating resistin level, while aerobic/endurance exercise increases the level in healthy populations in particular. The summarized results and experimental/ clinical evidence indicate that resistin can have functions related to inflammation and oxidative stress. The circulating resistin level may be also useful as a marker of the effects of exercise. Further studies are needed.

Chapter 3 - Resistin is a peptide that belongs to resistin-like molecules family. In humans, it is mainly produced from macrophages, bone marrow cells, and peripheral blood mononuclear cells. Resistin promotes inflammation via up-regulation of inflammatory cytokines' production and secretion especially in some proinflammatory conditions, such as obesity, metabolic syndrome and type 2 diabetes mellitus (T2DM). Furthermore, chronic hyperresistinemiais strongly is associated with inflammation, which is the key factor in T2DM pathogenesis. By the inflammation favoring, the increased resistin levels worsen endothelial dysfunction and promote earlier onset of diabetes vascular complications. Also, resistin can promote vasculopathy through the promotion of macrophage scavenger receptors and CD36 atheroma macrophages expression, the formation of foam cells, as well as proliferation and migration of vascular smooth muscle cells. This favours endothelial dysfunction, which results in limited nitric oxide and increased superoxide production. Some studies revealed the correlation of higher resistin levels with proatherogenic lipids, such as very low-density lipoproteins, Apo B, triglycerides, and small dense low-density lipoproteins, especially in T2DM patients. In this chapter, the authors have presented recent literature data related to the link between the resistin and diabetes.

Chapter 4 - Resistin is a 12.5 kD cysteine-rich protein belonging to the family of proteins named FIZZ (found in inflammatory zone) or RELM (resistin-like molecules). Resistin has an established function in mice as a molecular link between obesity and diabetes and an unclear function in humans, were it is presumed to be primarily involved in inflammation. In mice, resistin is produced by adipocytes, while in humans it is produced by neutrophils, monocytes, and macrophages. The human resistin gene is homologous by ~60% with mice resistin, explaining the differences in properties of the protein between rodents and humans. Resistin in humans is known to be expressed in atherosclerotic plaque, and in aneurisms and its plasma levels are associated with coronary heart disease (CHD). Accumulating data suggest that resistin may be involved in the initiation, progression and rupture of atherosclerotic plaque. Recent data reported that resistin may promote the initiation of atherogenesis by activating vascular endothelial cells (EC) and smooth muscle cells (SMC) and by interfering in their cross-talk, leading to an increase in vessel wall inflammation. Resistin, when present in the diseased vessel wall, may orchestrate a cascade of pro-atherosclerotic events, contributing to atherosclerosis and progression towards plaque rupture. Thus, resistin may augment monocyte and macrophage infiltration, may interfere in cell cross-talk, and may modulate the expression of various inflammatory cytokines, leading to acceleration of plaque growth and destabilization by mechanisms that are now being uncovered. Existence of a specific resistin receptor in immune and vascular cells is controversial, but recent data indicating that the Toll-like receptor 4 (TLR4) and Gi protein signalling are involved in resistin effects in vascular cells and that adenylyl cyclase-associated protein 1 (CAP1) may function as a receptor for human resistin. Resistin and fractalkine, which is an atypical chemokine that functions as chemoattractant or as adhesion molecule, were found at increased levels in human atherosclerotic plaque, in the sub-endothelium, resistin being associated with CD68 monocytes/macrophages and fractalkine predominantly associated with the intimal SMC and with monocytes/macrophages. Recent data uncovered fractalkine as a new molecular link between resistin and inflammation. This chapter reviews resistin's role in vascular inflammation in atherosclerosis, the molecules and mechanisms involved, and their potential use as new targets for innovating nanobiotherapies.

Chapter 5 – Introduction. This chapter deals with an important issue, based on the known sequential events of obesity preceding severe acute pancreatitis (AP). The task of investigating a potential new predictive marker of fat necrosis and pancreatitis severity will provide a useful clinical tool. This novel adipose-driven protein (Resistin), produced by pancreatic islets, may represent a useful additional marker for the severity of AP. *Methods:* This prospective study was conducted over three consecutive years. The authors included 112 patients who were diagnosed with acute biliary pancreatitis. The authors measured body mass index (BMI), waist circumference (WC), and serum resistin levels by enzyme-linked immunosorbent assay (ELISA) and compared this with the control group having similar gender, BMI, and age. *Results:* A total of 112 patients were diagnosed with acute gallstone-induced pancreatitis. The mean BMI value was 31.4 kg/m2 (obese, class 1). Twenty seven patients (24.1%) had severe AP, while eighty-five patients had mild pancreatitis (75.9%). The authors' results showed that the BMI significantly correlated with pancreatitis severity ($P = 0.007$). Serum resistin levels were significantly higher in patients with AP when compared with control subjects ($P < 0.0001$). *Conclusion:* It was demonstrated that there is a correlation between serum resistin and AP, and therefore it can be used as a diagnostic marker for AP.

In: Resistin
Editor: Gerald Maldonado
ISBN: 978-1-53614-543-4


*Chapter 1*

# RESISTIN: STRUCTURE, FUNCTION, AND ITS ROLE IN THE PATHOPHYSIOLOGY OF OBESITY, DIABETES, METABOLIC SYNDROME, CARDIOVASCULAR DISEASES, AND THE NERVOUS SYSTEM

***Sanya Haiaty[1], Zahra Barabadi[3], Mostafa Mostafazadeh[4,2], Ali Rastqar[5,6], Mahtab Keshvari[7,*] and Sedigheh Asgary[7,†]***

[1]Infectious and Tropical Diseases Research Center, Department of Clinical Biochemistry, Tabriz University of Medical Sciences, Tabriz, Iran
[2]Department of Biochemistry and Clinical Laboratories, Tabriz University of Medical Sciences, Tabriz, Iran
[3]School of Medicine, Sabzevar University of Medical Sciences, Sabzevar, Iran

[*] Corresponding Author Email: Mahtabkeshvari87@yahoo.com.
[†] Corresponding Author Email: sedighehasgary@gmail.com.

[4]Immunology Research Center, Faculty of Medicine,
Tabriz University of Medical Sciences, Tabriz, Iran
[5]Department de Psychiatry et Neuroscience,
Université Laval, Québec, QC, Canada
[6]Centre Hospitalier de l'Université Laval (CHUL),
Québec, QC, Canada
[7]Isfahan Cardiovascular Research Center, Cardiovascular Research
Institute, Isfahan University of Medical Sciences, Isfahan, Iran

## ABSTRACT

Over the past decade, adipose tissue has received much attention not only for its endocrine properties but equally also for its immunological activity. In consideration of this, it has become increasingly apparent that adipose-secreted factors such as leptin, tumor necrosis factor (TNF)-a, IL-6, adiponectin and more recently, resistin appears implicated in some inflammatory and metabolic disease. Resistin, in particular, has received meticulous attention as a potential factor in obesity-mediated type 2 diabetes mellitus (T2DM), Metabolic Syndrome, and Cardiovascular. Serum levels of resistin, an adipose tissue-derived adipokine, are increased in human obesity and are positively correlated with atherosclerotic cardiovascular disease. As such, several notable questions remain to be answered within the field of resistin biology. For instance, what conclusion can be drawn with regards the physiological function of resistin? Do the studies over the past year evaluate a role for resistin in the pathogenesis of obesity-associated inflammation and T2DM, Metabolic Syndrome, and Cardiovascular, or merely add to the chaos and confusion in the existing literature?

These chapter results reveal that resistin is a highly attractive therapeutic target in factors related to diseases humans. Because current evidence appears to suggest that resistin is a proinflammatory cytokine. Thus, like many other adipocytokines, resistin may possess a dual role in contributing to metabolic disease: first through its direct effects on substrate metabolism and second, through regulating inflammation within its target tissues. The chemistry of resistin has also been the subject of investigation and like adiponectin, the homo-oligomerization of this protein has a bearing on its function.

This chapter reviewed to summarize data about structure and function resistin. Also, role and contribution of resistin in the initiation and progression of pathophysiology actions among human and other mammalians reviewed. Summarize data have obtained from clinical researches and animal studies.

## 1. Structural Biochemistry and Expression of Resistin

Resistin (or 'resistance to insulin'), also known as adipocyte-secreted factor (ADSF) [1], C/EBP-epsilon-regulated myeloid-specific secreted cysteine-rich protein (XCP1) or found in inflammatory zone 3 (FIZZ3) [2], was discovered in 2001 by Steppan et al. to describe a small circulating mouse protein that was specifically expressed and secreted by adipocytes and the serum levels of which increased markedly in mouse models of genetic and diet-induced obesity [3].

Resistin is a secretory protein that belongs to the resistin like molecules (RELMs), a small family of mammalian secreted proteins which encoded by RELM genes (Retn) family and related with the activation of inflammatory processes, with different expression patterns and biological activity [4]. RELM genes encode secretory proteins of 105 to 138 amino acids (aa) with three essential domains: an amino-terminal signal sequence, a variable centric section, and a preserved carboxyl-terminal (C). The carboxyl-terminal is comprised of a 10–11-cysteine-rich motif sequence shared by all RELM family proteins which sustain the globular domain of the resistin monomer through the formation of five disulfide bonds. This C terminal domain has been proposed to plays an important role in the binding of resistin to its receptor [5, 6]. The Resistin like molecules family is made of RELM-α, RELM-β, and RELM γ. RELM-α, a secreted protein with its highest levels of expression in adipose tissue and also in allergic inflammation, is the only monomeric member of RELM family. RELM-β is found in the gastrointestinal tract and has been identified as a possible source to tumors in the colon. RELM-β is most similar in structure to resistin as both are disulfide dependent homodimers. RELM γ has only recently been identified and is most highly expressed in the hematopoietic system, with high expression in the bone marrow, and lower expression in white blood cells, spleen and thymus [4].

Resistin is expressed in several cell types including adipocytes [3, 7], stomach and intestinal epithelium, suprarenal gland cells, myocytes [8] testis [9], and possibly astrocytes [10]. Resistin in rodents, mainly secreted by white adipose tissue, is an 11 kDa cysteine-rich polypeptide which its gene (RETN) is located on chromosome 8A1. It is synthesized as a 114 aa precursor, with a signal peptide of 20 aa and a 94 aa mature segment. Amino acid identity in the mature segment within mice and rats is 72% [3, 11]. Unlike rodents, in humans resistin is mainly produced by tissue macrophages, peripheral blood mononuclear cells (PBMCs), spleen, and bone marrow cells [12, 13]. The human resistin gene, located on chromosome 19p13.3, encodes a 12.5 kDa cysteine-rich polypeptide with a mature segment consisting of 108 amino acids. Indeed, the human and rodent resistin genes have significantly different promoter regions, suggesting various mechanisms of regulation, tissue distribution and activities [3]. The comparison of the amino acid sequence of cow resistin with that of human, pig, rat, and mice revealed 73, 80, 58 and 57 percent homology respectively [14].

The resistin protein is constructed from 3 monomeric chains, A, B, and C, which are dimerized with another resistin molecule (trimeric) to formation its biologically active form. Each chain contains 94 amino acids and only Chain A is sequence unique. The secondary structure of chain A contains 1 α-helix of 22 residues (23% helical) and 6 β strands composed of 39 residues (41% beta sheet). Chain B is 24% helical (23 aa) and 40% beta sheet (38 aa) and Chain C is 23% helical (22 aa) and 42% beta sheet (40 aa). The isoelectric point (pI) of resistin protein is 8.01. Each chain of resistin trimeric molecule (A, B, and C) have two domains: the amino terminus alpha helical "tail" and the carboxyl terminus disulfide rich beta strand "head" (5 disulfide bonds, Cys35-88, Cys47-87, Cys56-73, Cys58-75, and Cys62-77). This "head-tail" conformation is dissimilar to other proteins found in the Protein Data Bank (PDB) (except for RELM-β), but the general structure of the head is resembling octopus hemocyanin [15].

Normally, the circulating levels of resistin range from 7 to 22 and 36–43 ng/ml in human and rats and its concentrations are significantly higher in females vs. males.

The follicular fluid resistin concentrations in humans and pig are from 5 to 50 [16] and 0.32 ng/mL respectively, dependent on the stage of the estrous cycle [17]. It has been shown that in obese and diabetic patients, the circulating level of resistin is significantly increased [3, 18]. Resistin expression in rodent adipocytes is stimulated by high levels of glucose but suppressed by tumor necrosis factor α (TNF-α) while human resistin is strongly induced by various inflammatory stimuli, including interleukin 6 (IL-6), IL-1β, lipopolysaccharide (LPS), TNF-α, and resistin itself, in PBMCs and macrophages [19].

## 1.1. Resistin Receptors and Its Signal Transduction

According to data have been obtained from animal and human studies, the receptor or receptors of resistin especially in human is not fully understood. But, Toll-like receptor 4 (TLR4) and adenylyl cyclase-associated protein 1 (CAP1) are accepted as putative candidates for its receptors [20]. TRL family is expressed in various organisms including mammalians. It has been reported 13 TLR (TLR1-13) so that, TLR1 to 9 are both in human and mice, TRL10 is not functional in mice but in human is active, and the other ones do not express in human but only in mice [21-23]. TLR1-2, 4, 5-6, and 10 present on the cell surface, whereas the others are endosomal or lysosomal receptors [21, 22].

TLRs essentially participate in immunologic pathways and are divided into some groups while each group has own signaling pathway (for more detail, see ref [22]). Among TLRs, TLR4 is proposed one of resistin membrane receptors that mediates its intracellular functions [24, 25]. In fact, TLR4 is considered as an immune receptor that promotes the production of inflammatory substances in macrophages but for example, it can also mediate the effects of resistin in hypothalamus for activating pro-inflammatory pathways [21, 25].

TLRs are found in macrophages, neutrophils, dendritic cells, natural killer cells, and mast cells of the innate immune system; in T- and B-lymphocytes of the adaptive-immune system; as well as in some non-immune cells, such as epithelial and endothelial cells [26]. According to some evidence, among TLRs, TLR4 is unique because, after binding of an appropriate ligand, this receptor can undergo by endocytosis and follow its action as an internal receptor [27].

Up to now, they are found many endogenous ligands can activate TLR4. These are including HSP22, EDA fibronectin, low molecular weight fragments of hyaluronic acid, fibrinogen, and tenascin-C whereas, lipoteichoic acid, LPS, CpG motifs of bacterial DNA, and viral RNA are exogenous ligands [22, 28]. After binding the ligands, TLR4 usually exert its effects via two classic models of signaling pathways. They are MyD88 (myeloid differentiation primary response gene 88) and Toll-interleukin 1 receptor domain-containing adaptor inducing interferon-beta (TRIF) pathways. Then, autophosphorylation of interleukin-1 receptor-associated kinases (IRAK) and oligomerization of TNF receptor-associated factor 6 (TRAF6) is occurred and cause the activation of IKK and mitogen-activated protein kinase (MAPK). Finally, subsequent signal propagation results in the translocation of nuclear factor kappa B (NF-κB) to the nucleus and the activation of the activator protein-1 (AP-1) transcriptional program [21, 22, 26, 28].

On the other hand, downstream of TRIF, in addition to the delayed activation of NF-κB via an interaction with TRAF6, the ubiquitination of TNF receptor-associated factor 3 (TRAF3) induces TANK-binding kinase 1 (TBK1) binding to IκB (inhibitor of NF-κB) kinase epsilon (IKKε). Then, the TBK1-IKKε complex phosphorylates the transcription factor interferon regulatory factor 3 (IRF3), ultimately driving expression of interferon beta (IFNβ) and interferon-inducible protein 10 (IP-10). See Figure 1 for a better understanding of the details [29].

The other receptor, which has proposed as a putative resistin receptor, is CAP1. Although TLR4 has been extensively studied and even there is a report that revealed that resistin in mice binds to tyrosine kinase-like orphan receptor 1 (ROR1) [30, 31], it can directly bind to CAP1 in human monocytes [31]. During the evolution, CAP1 is conserved from yeast to human. It is an action-binding protein that regulates actin cytoskeleton and cell adhesion [31, 32]. See Figure 2 for a better understanding [29].

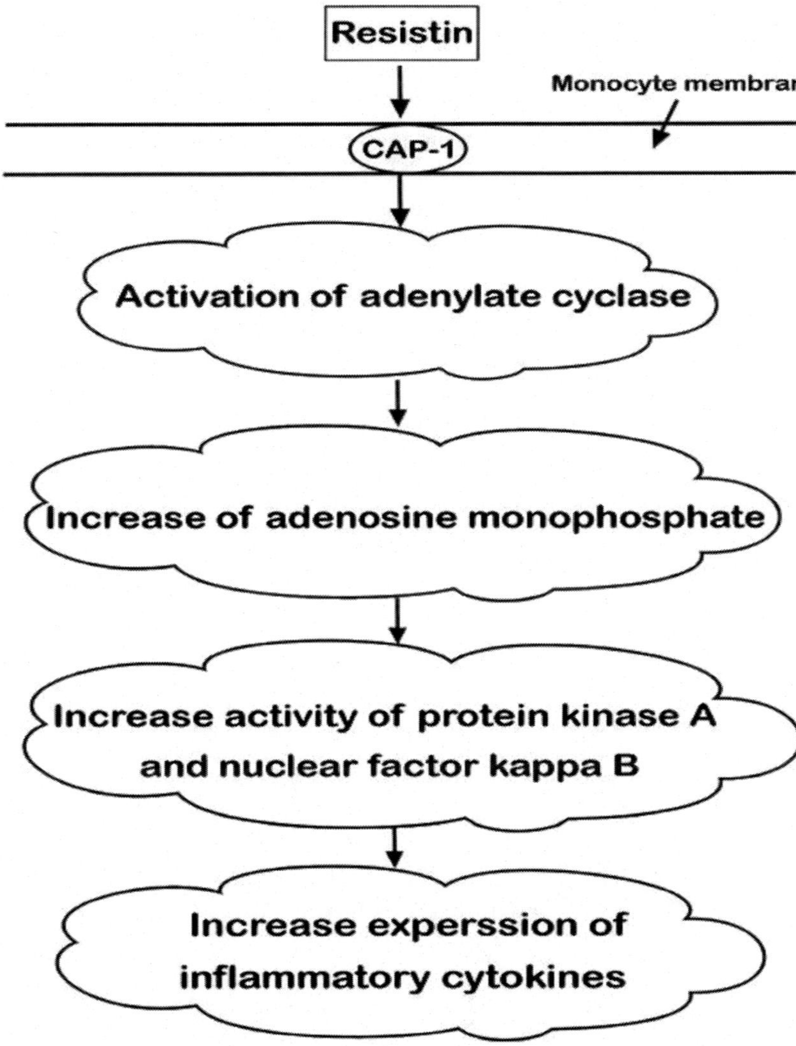


Figure 1. Tool-Like Receptor 4 (TLR4). A suggested mechanism for resistin cytoplasmic and nuclear signal transduction by TLR4 in vertebrates [33].

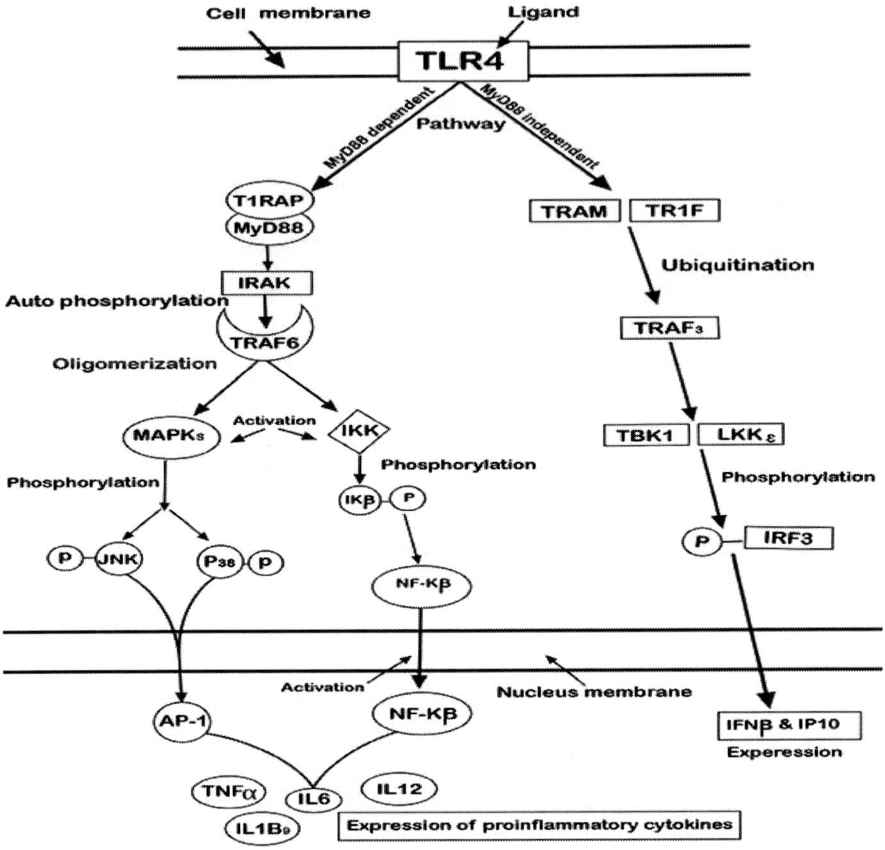


Figure 2. Adenylyl cyclase-associated protein 1 (CAP1). Intracellular signaling of resistin via CAP1 in human monocytes [33].

## 1.2. Resistin and Inflammation (Proinflammatory Cytokines, Tumor Necrosis Factor Alpha, Interleukin-6 and Others)

The resistin molecule plays an important regulatory role in the inflammatory response during which macrophages, PBMC, and endothelial cells are the primary targets of resistin. The regulatory role of resistin in inflammatory responses has been clearly demonstrated. During the inflammatory reaction, macrophages, peripheral blood mononuclear cells and endothelial cells are the primary targets of resistin [20]. *In vitro* and *in vivo* studies on human monocyte culture and human resistin-expressing transgenic mice respectively, propose that resistin is a stimulatory cytokine which promotes immune cell activation and chemotaxis, neutrophil extracellular traps (NETs) formation and production of inflammatory cytokines [31, 34]. It has been demonstrated that resistin induces the mRNA expression of 20 tested cytokines and chemokines including Tumor necrosis factor alpha (TNF-α), Interleukin-1α (IL-1α), IL-1β, IL-6, IL-12, CCL2, CCL3, CCL3L1, CCL4, CCL5, CCL8, CXCL1, CXCL2, CXCL3 in normal human chondrocytes [35] and monocyte chemoattractant protein (MCP)-1 in PBMCs, macrophages, and hepatic stellate cells through the nuclear factor-κB (NF-κB) pathway [36]. Also, a recent study demonstrated that resistin promotes the production of pro-inflammatory cytokine (IL-1β, TNF-α, and IL-6) in porcine alveolar macrophages via TLR4/NF-κB-mediated pathway [37]. A significant association between plasma resistin levels and TNF-α, IL-6, IL-12, MIP-1α, MIP-1β, and CD40L has been shown in patients with atherosclerosis [38]. Furthermore, another study demonstrated that resistin induces the production of the TNF-α and IL-6 cytokines via the NF-κB-dependent pathway during acute pancreatitis [39]. Moreover, serum levels of resistin are correlated with inflammatory markers such as C-reactive protein (CRP), TNF-α, and IL-6 in the individuals with endotoxemia, diabetes mellitus type 2 (T2DM), coronary artery disease (CAD), inflammatory bowel disease (IBD), chronic kidney disease (CKD), rheumatoid arthritis (RA), and/or sepsis [20]. Plasma resistin levels are also correlated with disease severity in patients with acute pancreatitis or severe sepsis and are predictive of poor outcomes in non-septic but critically ill patients [19].

## 1.3. Resistin Expression and Action at Hypothalamus–Pituitary Level

Although adipose tissue is the more relevant source of resistin, RT-PCR studies have been revealed that resistin is also expressed in the hypothalamus of rodent [10, 40]. In addition to RT-PCR studies, immunohistochemical analysis demonstrated the presence of resistin protein in the mouse brain where it is located in the cells of the arcuate nucleus. A few cells in the ventrolateral and ventromedial nuclei of the hypothalamus and in the dorsal periventricular region of the mouse also contain resistin. Furthermore, Gene expression studies also revealed that resistin mRNA expression is considerably higher in the arcuate nucleus compared to other parts of the hypothalamus. These findings suggest that resistin probably produced and secreted by the arcuate nucleus in the hypothalamus [10]. Evaluation of the resistin mRNA expression by RT-PCR assay indicated that resistin is also expressed in mouse pituitary gland [40, 41]. Morash and et al. also showed that the resistin mRNA is mainly localized to the anterior and intermediate lobes, with a little amount in the posterior lobe of pituitary gland. These results also have been confirmed by immunohistochemistry (IHC) analysis [10]. In the pituitary, gender differences of resistin are evident (male > female), resistin expression is age-dependent and the expression of resistin mRNA is significantly higher in prepubertal mice [10, 40]. Moreover, corticosteroids (e.g., dexamethasone) significantly increased the expression of resistin mRNA in the pituitary gland [42].

It has been shown that the treatment of rat pituitary cell cultures with resistin increases the release of the growth hormone, reduces the expression of the adipokine receptor (AdipoRs), and increases the intracellular calcium [$Ca2^{+}$] in single somatotropes. These results together implicate the existence of resistin receptors on the pituitary somatotropes, a receptor which it has not been cloned yet [43]. Recently, it has been demonstrated that central application of resistin regulates lipid metabolism in hypothalamic and peripheral tissues in a nutritional-dependent manner. Also, it has been shown that the anorectic effect of resistin is related with the low levels of mRNA expression of the agouti-related protein and neuropeptide Y and the increased mRNA expression of cocaine and amphetamine-regulated transcript in the arcuate nucleus of the hypothalamus. Also, it has been shown that resistin exerts a nutritional status dependent inhibitory effect on lipid hemostasis in hypothalamus. resistin increases the phosphorylation levels of both AMP-activated protein kinase (AMPK) and its downstream target acetyl-coenzyme A carboxylase, associated with decreased expression of fatty acid synthase in the hypothalamus[44].

## 2. Resistin as a Potential Pharmacological Target

Several clinical studies have demonstrated that serum resistin levels may serve as an informative marker for metabolic disease in humans, particularly type 2 diabetes, myocardial infarction, and atherosclerosis [45-49]. Also, clinical evidence supports the role of resistin as a potential marker of Inflammation, Tumorigenesis, Rheumatic diseases, Osteoporosis, Chronic kidney disease, Asthma [50]. Therapy targeting the reduction of serum resistin levels is a promising strategy for coronary artery disease, as well as other diseases [51]. Therapeutic intervention for treating these diseases alter several signaling pathways related to inflammation [51], glucose metabolism [50], Lipid Metabolism [52].

Resistin was identified as a hormone with effects on glucose metabolism which may contribute to insulin resistance [53]. On the other hand, glucose-lowering drugs that used for treating type 2 diabetic patients have regulated resistin expression [50, 54].

The conflicting results on resistin expression were accumulated after treatment with antidiabetic agents [52, 55].

Glitazones are anti-diabetic drugs that antagonizing insulin action with regulating resistin expression [54]. Rosiglitazone and other glitazones are a peroxisome proliferator-activated gamma (PPAR-g) receptor agonist that reduced glucose and lipid levels in patients with type 2 diabetes and led to reducing serum resistin concentration in mice. However, the following studies confirmed reduced resistin expression in adipose tissue of obese animals, [54] and from obese humans [56, 57].

Thiazolidinediones (TZDs) decreased serum resistin levels in both mice and humans and inhibited the expression of resistin in cultured 3T3-L1 adipocytes and in white adipose tissue in db/db mice. Surprisingly, another study showed that treatment with TZDs increased resistin expression [50, 52, 55]. Metformin is another oral glucose-lowering drug which is widely used for treating type 2 diabetic patients. Also, it is known to improve hyperglycemia and insulin resistance by reducing gluconeogenesis and by enhancing peripheral glucose uptake [58-61]. In addition, metformin treatment causes weight loss by decreasing adipose tissue in obese patients with type 2 diabetes [55, 60].

However, the outcome of various studies unexpectedly showed that resistin protein expression in adipose tissue was increased by metformin treatment [52, 55].

As well as the role of resistin in glucose metabolism, resistin has been also involved in the control of lipid metabolism [52]. Elevated serum resistin levels are involved in the myocardial infarction, and atherosclerosis [59] and are correlated with markers of inflammation and are prognostic of coronary atherosclerosis in humans, independent of plasma C – reactive protein [62]. Statins (atorvastatin and simvastatin) are HMG-CoA reductase inhibitors that inhibited a critical enzyme in cholesterol production in the liver while also having significant anti-inflammatory effects [51] and preventing of atherosclerosis progression [63]. Resistin, a novel adipokine, was originally proposed as a link between metabolic signals, inflammation, and atherosclerosis [62, 64]. Several studies indicated that atorvastatin decreases resistin expression [51, 65-67].

Shyu et al. have reported that TNF-α could induce resistin expression in human macrophages and atorvastatin was able to inhibit the resistin expression induced by TNF-α. The inhibitory effect of atorvastatin on TNF-α-induced resistin expression was mediated through race and resistin promoter [67].

Hu et al. have also reported that CRP induces resistin mRNA expression in human PBMC; co-incubation with simvastatin significantly inhibited this CRP-induced up-regulation of mRNA and protein expression of resistin [65].

Ichida et al. suggested that treatment with atorvastatin reduced resistin mRNA levels in 3T3-L1 adipocytes and human monocytes/macrophages [66].

Consequently, statin therapy could be a novel therapeutic strategy for regulating resistin associated cardiovascular dysfunction in humans [51].

According to the data mentioned above resistin play an important role in inflammation. Resistin is a proinflammatory cytokine that involved in inflammatory disorders; rheumatoid arthritis (RA) and inflammatory bowel disease [50, 68]. In patients with RA, high levels of resistin have been found in synovial fluid [69].

Also, more recent investigations have shown that a positive correlation between C-reactive protein (CRP) and erythrocyte sedimentation rate (ESR) with serum resistin in RA patients [70, 71].

TNFα blockers are highly effective in the treatment of RA [72, 73]. Infliximab and adalimumab are anti-TNF-α antagonist-monoclonal antibody that reduced serum resistin levels in patients with RA and inflammatory bowel disease [50, 51, 74].

As a result, serum-resistin levels were markedly decreased after anti-TNF-α antagonist-monoclonal antibody treatment in patients with RA [74]. Future studies will show if resistin can be used as a clinical marker for monitoring inflammatory changes in patients. Elevated serum resistin levels are involved in hypertensive patients. Clinical studies indicated that anti-hypertensive drugs have an effect on plasma adipocytokine levels (resistin).

Amlodipine is a calcium channel blocker that decreased plasma resistin levels in patients with arterial hypertension. As well as ramipril (an angiotensin-converting enzyme inhibitor) and candesartan (angiotensin II type 1 receptor blockade) significantly increased plasma resistin levels [75, 76].

Resistin participated in oxidative processes and regulated of murine adipokine gene expression [77]. Also, oxidative processes are involved in various diseases [78-82].

The antioxidant supplementation significantly reduced resistin serum levels [77, 83]. Several clinical studies suggested that the resistin expression and levels significantly affected by antioxidants supplementation: Retinoic acid, the acid form of vitamin A, inhibited the expression of resistin and reduced the higher resistin levels in mice [84], the antioxidant seleno-methionine, which increases the activity of glutathione peroxidase in endothelial cells, completely blocks vasomotor dysfunction induced by resistin in porcine arteries [85], oral folic acid consumption reduced plasma resistin level [83], vitamin C significantly reduced the plasma resistin level, with an increased level of reduced glutathione (GSH) [77]. Consequently, the treatment with antioxidant supplementation could be a novel therapeutic strategy for reducing resistin serum levels [77, 83].

# 3. Adiponectin-Resistance in Obesity Type 2 Diabetes

## 3.1. The Role of Resistin in Rodent's Obesity

Resistin in rodents is primarily expressed in and released from white adipose tissue (WAT) with some secondary expression from the pancreas, hypophysis, hypothalamus, gastrointestinal tract, suprarenal glands, and skeletal muscle. *In vitro* studies have shown that resistin expression follows adipocyte differentiation [1, 86], also in high-fat-fed mice, increased circulating levels of resistin is positively correlated with the adipocyte differentiation markers like fatty acid binding protein (AP-2) [87]. In rodent 3T3-L1 cells and primary pre-adipocytes, resistin expression is induced during the process of differentiation and maturation. The administration of exogenous recombinant resistin increases lipogenesis in 3T3-L1 cells [88], while resistin knockdown considerably decreases lipid content in 3T3-L1 cells after adipocyte differentiation, suggesting that resistin induces lipogenesis and inactivation of fatty acid β-oxidation in differentiated adipocytes [89].

Another study on murine C2C12 cells have been indicated that treatment with resistin leads to a reduction in expression of the fatty acid oxidation (FAO) transcriptional regulatory genes ERRα and NOR1, the fatty acid transport gene CD36, and the mitochondrial β-oxidation genes CPT1, MCAD, and ACO, which led to a reduction in the rate of cellular fatty acid oxidation [90].

Furthermore, overexpression of a dominant inhibitory form of resistin (Resistin-hFc, by fusing resistin to the human IgGγ constant region or hFc) blocks the resistin inhibition of adipocyte differentiation of 3T3-L1 cells. Also in transgenic mice, overexpression of Resistin-hFc in adipose tissue increased adipogenesis and expression of adipocyte markers and promotes obesity [91]. As mentioned above, resistin is expressed in the hypothalamus, where it regulates hypothalamic and peripheral lipid metabolism and exhibit a nutritional-dependent anorectic effect [92, 93]. In this regard, experimental central application of resistin extremely prevents feeding and reduces bodyweight by regulating glucose and lipid metabolism [94].

It has been shown that plasma levels of resistin, elevate in genetically modified or high-fat-diet induced obese mice, a decline in fasting conditions, and quickly increase upon renourishing [1, 3, 95]. To validate these findings, Lee and et al. showed that circulating levels of resistin is positively associated with body weight in various mouse models of obesity (i.e., high-fat-fed obese wild-type mice, diet-induced obese (DIO) C57BL/6 J mice, high-fat-fed obese TNF-α−/−, and brown adipose tissue-deficient uncoupling protein–diphtheria toxin A chain (UCP1-DTA) mice [96]. In addition, resistin mRNA level in rat adipose tissue is positively correlated with stomach contents and serum levels of insulin [97]. Also, feeding rats with a high-fat diet combined with fructose in drinking water, which mimics a high-fat diet/fructose western diet, increased the serum levels of resistin [98], and rats fed a high-carbohydrate regimen comprising of refined sugar exhibit a quick and transient increase in plasma levels of resistin. After a chronic stress, feeding mice with a high fat diet has led to further increases in plasma levels of resistin [99]. After weight cycling, serum levels of resistin declines during standard chow feeding and elevates during high-fat feeding and is positively correlated with alterations in the intraabdominal and hypodermic fat pads [100]. Some recent studies have shown contradictory results regarding the relationship between obesity and expression of resistin mRNA in adipose tissue. Contrary to expectation, decreased the expression of resistin mRNA in the adipose tissue or isolated adipocytes has been reported in different genetic models of obesity [101-104]. Altogether, the findings heretofore indicate that plasma levels of resistance in rodents are positively associated with obesity and a high fat/high carbohydrate diet. From these combined findings, researchers have come to the conclusion that resistin acts as both a negative regulator of adipogenesis and as a fat sensor of nutritional status [105]. Based on the above evidence, increased levels of resistin are commonly observed in feeding rodents with a high fat/high-carbohydrate diet, obese conditions physiologically characterized by insulin resistance and impaired glucose uptake.

## 3.2. Resistin in Glucose Hemostasis in Rodents

It has been shown that resistin plays an important role in the progress of insulin resistance in rodents. *In vitro* administration of resistin to murine 3T3-L1 cells (adipocyte cell type), cardiac and skeletal myocytes reduced insulin-induced glucose uptake into the cells [3, 106, 107]. It has also been shown that *in vivo* administration of resistin impairs insulin function in normal mice [3]. Central and peripheral application of recombinant resistin or transgenic overexpression of resistin in mice leads to hepatic insulin resistance [3, 95, 108]. An experimental study have been showed that *in vitro* administration of resistin decrease insulin secretion from INS-1E cells (rat pancreas cell) and pancreatic islets at normal and high concentrations of glucose, and also reduce glucagon secretion from G9 cells (Hamster insulinoma cell) and pancreatic islets at 1 mM, whereas it stimulates glucagon secretion at 6 mM glucose [109].

Resistin also promotes insulin resistance both in primary cultures of mice hepatocytes and in the liver of a mouse model with high liver-specific resistin expression [110]. Conversely, the neutralization of resistin with an anti-resistin IgG improves insulin sensitivity in diet-induced obese mice [3], and knockdown or deletion of resistin increases hepatic insulin sensitivity in high-fat-fed mice (as well as insulin sensitivity in muscle and white adipocytes in ob/ob mice), leading to a decrease in glucose production and an increase in peripheral glucose uptake [111]. After fasting, resistin knockout mice show lower glucose levels associated with a decrease in the expression of hepatic gluconeogenic enzymes [111].

As mentioned above, previous studies proposed a tyrosine kinase-like orphan receptor 1 (ROR1) as functional resistin receptors which may modulate white adipose tissue expansion or regulate glucose metabolism in rodents [88, 112]. It has been shown that murine resistin decreases the phosphorylation of AMP-activated kinase (AMPK) in hepatocytes, myocytes, and adipocytes [95, 113, 114] and the inhibition of resistin increased the AMPK phosphorylation, which is involved in glucose metabolism.

### 3.3. Human Resistin in Obesity, Insulin Resistance and Diabetes

Unlike rodents, the role of resistin in obesity and induction of insulin resistance in humans remains controversial. Accordingly, obese individuals who exhibit greater macrophagic infiltration from adipose tissue, compared to lean subjects, show increased expression of resistin in adipose tissue [93, 115]. Contrary to rodent white adipose tissue (WAT), the expression of resistin mRNA is very low in isolated human WAT [116-118]. The poor expression of resistin in human adipocyte can be due to the loss of a PPARγ genomic binding site, which controls the adipocyte-specific expression of the mouse Retn gene [119]. Notwithstanding these inter-species differences, it has been shown that human macrophage-derived resistin induces adipose insulin resistance in rodent [120].

As expected from animal study results, one primary study in healthy persons showed a strong association between fasting circulating levels of resistin and body fat percent [121], and two later studies indicated a considerable rise in serum resistin in obese subjects [121, 122]. On the contrary, plasma levels of resistin are significantly reduced in severely undernourished anorexic women and are declined following the hypocaloric diet- and exercise-induced weight loss [123]. In addition, weight loss in obese people is also associated with a decrease in circulating resistin levels, which correlated with changes in waist circumference (but not with changes in BMI). In one cross-sectional study, it has been shown that plasma levels of resistin are correlated with visceral, intrathoracic, and pericardial fat [124]. In another study, which conducted on children aged 6–8 and 12–16 years, no significant correlations were found between serum resistin concentrations and weight or BMI, as well as, no significant differences in plasma resistin levels observed between obese, overweight, and normal-weight children; however, a positive correlation between circulating resistin levels and fat mass has been shown in 12–16-year-olds, especially in girls [124]. Conversely, a study on morbidly obese patients before and after gastric bypass surgery showed that resistin levels in morbidly obese patients were not correlated with adiposity anthropometric measures, insulin, glucose, HOMA-IR [125].

These combined results indicate that plasma resistin concentrations in humans are correlated with central and visceral obesity, but this relationship does not apply BMI. However, other human studies have shown conflicting results. A cross-sectional study showed that plasma levels of resistin did not correlate with BMI, waist-to-hip ratio, or fat mass [126]. There was no change in plasma resistin levels after weight loss in obese individuals with insulin resistance as well as, in non-diabetic morbidly obese patients, resistin levels did not change after massive weight loss following biliopancreatic diversion (BPD) or the *Scopinaro procedure* [127]. After induction of weight loss by laparoscopic adjustable gastric banding (LAGB) surgery, resistin serum levels increased after 6 months but fell below baseline levels after 12 months. [128]. Moreover, two-day and four week caloric limitation programs in healthy, normal-weight individuals were not correlated with changes in plasma resistin levels [129, 130]. Overall, these inconsistent results do not allow a deterministic correlation between serum resistin levels and obesity in humans at this time.

In spite of that adipocyte-derived resistin has been certainly related to the expansion of insulin resistance in rodents, contradictory results have been shown in prior human studies. Several previous studies have reported a significant association between insulin resistance and plasma levels of resistin in healthy [131-133] and type 2 diabetic [134] persons. However, several other studies have failed to identify significant associations between insulin resistance and circulating resistin levels [18, 135, 136]. A study showed that circulating resistin levels were significantly higher in type 2 diabetic patients as compared to those with impaired glucose tolerance and euglycemic persons [137]. In addition, a cross-sectional study on Spanish obese individuals indicated a significant positive association between serum levels of resistin and insulin sensitivity [69].

In contrast, a recent pilot study on insulin-sensitive obese and insulin-resistant obese subjects revealed that insulin resistance was not significantly related to serum resistin levels, but had a significant positive correlation with visceral fat [138]. A meta-analyses study has been reported that there is no association between gestational diabetes and plasma resistin levels [138]. Recently, it has been shown that in patients with T2DM, the main determinants of resistin level are the level of renal function (eGFR) and inflammation (hs-CRP) rather than the presence of microvascular complications, obesity and insulin resistance [139]. An observational study in China suggests that serum resistin is positively correlated with serum lipids, but not with insulin resistance, in first-degree relatives of type-2 diabetes patients [140]. Although a definitive role of resistin in obesity and insulin resistance in humans has not yet been found, but its role in the pro-inflammatory processes, which clearly linked to the progression of obesity and insulin resistance in humans, is clear and definitive [141]. *In vitro* expression of resistin in peripheral blood mononuclear cells (PBMCs) is increased by the pro-inflammatory cytokines (TNF-α & IL-6), in turn, resistin induces the expression of these factors [117, 142]. As mentioned earlier, it has been showed that circulating resistin levels is associate inflammatory markers such as soluble TNF-α receptor-2 (TNFR2), IL-6, CRP, and lipoprotein-associated phospholipase A2 (Lp-PLA2) [143, 144]. Given that chronic and low-grade inflammation is a key reason for obesity-induced insulin resistance and crosstalk between inflammation pathway and insulin signaling, circulating resistin may be a messenger between inflammation and insulin signal transduction [145, 146].

## 4. Role of Human Resistin in Atherosclerosis

Many epidemiological, clinical and genetic studies investigated the association between serum level of resistin and cardiovascular disease.

Resistin appears to be one of the potential biomarkers of cardio-metabolic diseases associated with poor prognosis in these patients. In human, resistin plays its role in atherosclerosis with affecting different types of cells in cardiovascular system such as vascular endothelial cells, vascular smooth muscle cells (VSMC), hepatocytes, platelets and PMNCs like macrophages.

Atheriogenesis has also been reported to be induced by some other members of resistin-like molecules (RELM) like RELMb, that contributes in the development of arteriosclerosis via impact on lipid profiling, accumulation of lipids through altering the metabolism of fatty acids, hypertension, inflammation, endothelial dysfunction and vascular remodeling [147].

### 4.1. Resistin and Endothelial Cells

In endothelial cells, resistin with raising the expression of adhesion molecules such as intercellular adhesion molecule-1, vascular cell adhesion molecule-1, and P-selectin as well as pro-inflammatory chemokines including MCP-1, endothelin-1, matrix metalloproteinases and pentraxin 3 [148-150], leads to adhesion of immune cells to vascular wall of coronary arteries by a mechanism involving P-selectin and fractalkine and through p38MAPK signaling pathway that promotes the infiltration of leukocytes via increasing the permeability of coronary vessels [148, 151], and by this way induces an inflammatory environment in local area. Moreover, oxidative stress in these cells inhibits endothelial nitric oxide synthase (NoS) and nitric oxide (NO) production, a strong vasodilator, that leads to vasoconstriction [152]. Furthermore, it increases the proliferation and migration of endothelial cells, that together with increasing the release of endothelin-1 from these cells, has been proposed as causes of restenosis after artery stenting in diabetic patients [153, 154].

## 4.2. Resistin and Vascular Smooth Muscle Cells

In vascular smooth muscle cells, resistin exerts it effects with increasing cell adhesion via induction of VCAM-1 expression, upregulates osteopontin expression via activator protein 1 (AP- 1) [149] and upregulates CX3CL1 and CX3CR1 expression through transcription factors NF-kB, AP- 1, and signal transducers and activators of transcription 1/3 (STAT1/3) signal pathways [149]. The proliferation of these cells increases in response to resistin in a dose-dependent manner via ERK1/2 and Akt signaling pathways, that is considered as a potential mechanism of development of cardiovascular diseases [155]. It has been proposed that a decrease of NO production in response to an increased level of resistin in serum might be another reason for this proliferation.

## 4.3. Resistin and Adipocytes

mRNA expression of resistin increases in response to an elevated level of homocysteine in adipose tissue that consequently, leads to vascular smooth muscle cells (VSMC) migration. It has been shown that co-culture of VSMC with adipocytes stimulates VSMC migration in culture, which is induced by cytoskeletal changes and a5b1- integrin activation through an a5b1-integrin-focal adhesion kinase/paxillin-Ras-related C3 botulinum toxin substrate 1 pathway [156].

An association between resistin and malondialdehyde which is a marker of lipid peroxidation also has been reported in the obese young population that indicates oxidative and nitrosative stress triggering role for resistin that could predict the risk of cardiovascular disease in the future [157].

### 4.4. Resistin and Platelets

In platelets, resistin, through a p38 mitogen-activated protein kinase (MAPK)-dependent pathway increases P-selectin expression that leads to platelet activation [158]. The effect of resistin on downstream effectors of phosphoinositide 3 kinase (PI3K) and MAPK pathways in platelets activation and aggregate formation also has been reported [159]. All data on the effect of resistin in platelets indicates the thrombogenic role of this molecule that underlies the pathogenesis of CVD.

### 4.5. Resistin and Hepatocyte

It has been observed in an in-vitro study that resistin in hepatocyte causes a decrease of LDL receptors (LDLR). This effect is mediated through lysosomal degradation of LDLR in these cells by enhanced expression of a kind of protease known as proprotein convertase subtilisin/kexin type 9 (PCSK9). Reduction in LDLR decreases the uptake of LDL in the liver with consequently lipid accumulation in plasma, which further confirms the atherogenic role of resistin. And in this study, it has also been shown that resistin inhibited upregulation of LDLR caused by statins [160]. Another proposed mechanism by which resistin can changes the plasma lipid profile is the effect on hepatocytes to increase the expression and secretion of VLDL apoB via increasing the activity of microsomal triglyceride transfer protein (MTP) that finally would lead to accumulation of fat molecules in hepatocytes, a condition named fatty liver [161].

### 4.6. Resistin and Macrophages

Macrophages are both, the main source of circulating resistin in human and also the main target for it. Human resistin can cause arteriosclerosis and cardiovascular disorders through chronic inflammation by targeting vascular endothelial cells and stimulating some types of immune cells like macrophages and PBMCs to secrete pro-inflammatory cytokines as well as affecting vascular smooth muscle cell proliferation, arterial inflammation, and the formation of foam cells.

Resistin induces foam cell formation in human macrophages through up-regulation of SR-A and CD36 scavenger receptors and downregulation of the reverse cholesterol transporter ABCA1 (ATP-binding cassette transporter A1) and induces a prothrombotic state in human endothelial cells [162, 163].

Given together, these findings indicate a strong role for resistin in the pathogenesis of the cardiovascular diseases through promoting arteriosclerosis and pro-inflammatory state in the cardiovascular system (Figure 3).

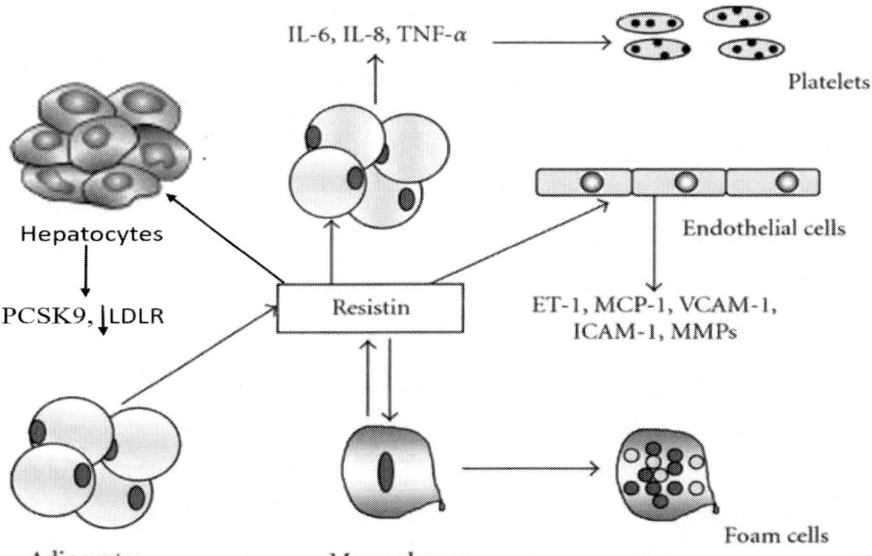


Figure 3. Potential vascular effects of resistin. IL-6, interleukin-6; IL-8, interleukin-8; TNF-α, tumor necrosis factor-α; ET-1, endothelin-1; MCP-1, monocyte chemoattractant protein-1; VCAM-1, Vascular Cell Adhesion Molecule-1; ICAM-1, Intercellular Adhesion Molecule-1; MMPs, matrix metalloproteinases. PCSK9, proprotein convertase subtilisin/kexin type 9; LDLR, LDL receptor [modified from 164].

## 4.7. Clinical Implications of Human Resistin in CVD

Many clinical studies have reported the association between elevated levels of resistin and coronary and peripheral artery diseases such as congestive heart failure (CHF), ischemic stroke, and myocardial infarction [165]. These conditions could be attributed partly, to atherosclerotic and thrombogenic properties of resistin. Restenosis after coronary artery stenting [115], the lower amputation-free survival rate in patients with diabetes undergoing lower extremity bypass surgery [165] and recurrence of coronary events [166], are some of these reported conditions associated with higher serum levels of resistin in compared to the normal level. Higher prevalence of cardiac failure and poor cardiac outcome in individuals with higher resistin levels have also been demonstrated. A strong correlation between the level of resistin and severity of CAD disorders has also been reported [167], whereas, some other studies have found no association between the serum level of resistin and cardiovascular diseases [168, 169].

The result extracted from a multi-ethnic cohort study showed a strong positive correlation between some cardiovascular risk factors such as body mass index (BMI), waist circumference, and systolic blood pressure and plasma level of resistin, but not sex [170].

## 4.8. Role of Human Resistin in Angiogenesis

Resistin is known to be able to induce proliferation and migration of human coronary artery endothelial cell (HCAECs) in culture. On Matrigel, it promotes tube-like structure formation and angiogenesis which is a pathologic state in neoplasm, ischemia and cardiovascular diseases [171]. Upregulation of some genes involved in angiogenesis including vascular endothelial growth factor receptors (VEGFR-1 and VEGFR-2) and matrix metalloproteinases (MMP-1 and MMP-2) has been identified. Resistin mediated this effect through phosphorylation of ERK1/2 and p38 in HCAECs [171]. In an in-vitro study, human umbilical vein endothelial cells treated with resistin showed tube like structure formation followed by vascular endothelial growth factor production [172]. In murine, resistin induces angiogenesis via phosphatidylinositol 3-kinase (PI3K)/AKT phosphorylation and NF-κB and β-catenin activation as well as VEGF production in murine epithelial cells [173].

## 4.9. Role of Human Resistin in Thrombosis

Thrombosis contributes to atherosclerosis development in metabolic syndrome (MetS). A significant correlation has been found between resistin level and tissue factor (TF) and plasminogen activator inhibitor-1 as markers of thrombosis in patients with MetS [174]. Treated HCAECs with resistin showed an increased expression of tissue factor (TF), a main characteristic of the pathophysiology of the acute coronary syndrome (ACS), by oxygen-free radicals and NF-κB activation, inducing a pro-thrombotic state. In an in-vitro study, expression levels of apolipoprotein C-I, angiotensin-converting enzyme, TNF receptor superfamily member 1A (TNFRSF1A) and CD40 was significantly increased in HUVEC cells treated with resistin, indicating that resistin exerts its thrombotic effects via lipoprotein metabolism [174]. Conversely, resistin with inhibiting eNOS and reduction of NO production counteracts with thrombosis [35, 175]. More studies are needed to clarify the role of resistin in thrombosis.

## 4.10. Role of Human Resistin in Hypertension

Many factors are known to be associated with hypertension. High level of resistin is suggested to be a risk factor for this medical condition. In some clinical studies, the relation between resistin level and hypertension has been demonstrated [176, 177]. High level of resistin detected in the young healthy offspring of individuals with essential hypertension, indicating the role of resistin as a prognostic factor in hypertension [178]. Atherogenesis, the proliferation of endothelial cells and VSMCs, reduced NO production, and increased endothelin production might be the underlying mechanisms by which resistin induces hypertension. However, more investigation would be necessary to clarify the precise mechanism of resistin regarding hypertension.

### 4.11. Role of Human Resistin in Inflammation

In human, the main source of circulating resistin are PBMCs rather than adipocytes in rodents. Its serum level is positively correlated with poor outcome in patients with sepsis and rheumatoid arthritis, and acute pancreatitis [115, 147]. High resistin protein level was also found in the area of inflammation in the patients with rheumatoid arthritis.

Although some receptors, like isoform of decorin, mouse receptor tyrosine kinase-like orphan receptor 1 (ROR1), and Toll-like receptor 4 (TLR4) are known as potential receptors for murine resistin, adenylyl cyclase-associated protein 1 (CAP1) seems to mediate inflammatory effects of resistin in humans. Activation of this receptor with human resistin leads to an increase in intracellular cAMP level followed by activation of both PKA and NF-kB pathways that finally leads to enhanced expression of inflammatory cytokines like interleukin-6 (IL6), tumor necrosis factor alpha (TNF-a), and IL-1b in monocytes that has been confirmed in both in-vivo and in-vitro studies [31].

On the other side, pro-inflammatory cytokines such as lipopolysaccharides (LPS), TNF-α, IL6 and IL1 have been demonstrated that are able to significantly increase resistin mRNA expression in human peripheral blood mononuclear cells [179].

Higher resistin mRNA expression and protein level were found in response to C-reactive protein (CRP) indicating the resistin involvement in inflammation [180]. Not only in PBMCs, but also in white fat cells in adipose tissue, resistin induces the production of pro-inflammatory cytokines such as IL-6, IL-8, and TNF-α that leads to an inflammatory state in this tissue [181].

# 5. ROLE OF HUMAN RESISTIN IN NEURAL DISEASES

Central nervous system (CNS) controls endocrine activities and is controlled by the endocrine system via hormones reciprocally. Among other hormones, adipokines including leptin, adiponectin, resistin, TNF-α, and IL-6 can easily pass blood-brain barrier (BBB) by simple diffusion or via transport systems and affect the CNS functions or alter some biochemical activities in the peripheral organs by modulating nervous system [182, 183]. It has proven that some of the adipokines such as leptin [184, 185] and adiponectin [186] are neuroprotective. Also, adipokines have the receptors in the CNS then they play a role in the physiology and pathophysiology of the CNS [187]. For summary, see Figure 4.

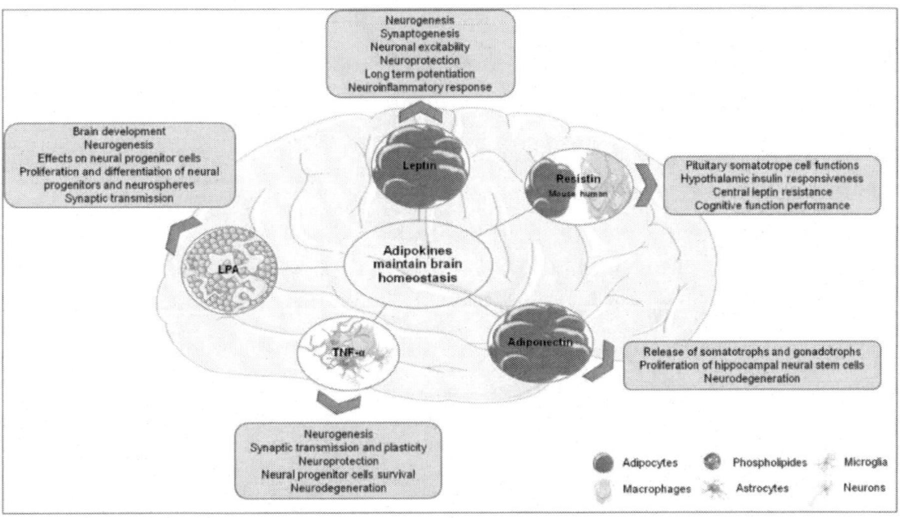


Figure 4. Effects of the main adipokines on brain homeostasis/functions. LPA: lisophosphatidic acid, TNFα: tumor necrosis factor-α [187].

Resistin (or adipose tissue-specific secretory factor: ADSF or C/EBP-epsilon-regulated myeloid-specific secreted cysteine-rich protein: XCP1) is known to play a key role in the CNS [187]. 6-hydroxydopamine (6-OHDA) is one of the well-known neurotoxins in Parkinson's disease (PD). In PD, there is a progressive degeneration of dopaminergic cells in the substantia nigra. Experimentally, 6-OHDA affects the mitochondrial respiratory chain and induces oxidative stress and dopaminergic neurodegeneration. In a study, resistin could abolish 6- OHDA-induced cell death in MES23.5 cells (a dopaminergic cell line hybridized from murine neuroblastoma-glioma N18TG2 cells with rat mesencephalic neurons) [182].

Central resistin administration could block Akt phosphorylation and enhance the expression of glucose-6-phosphatase (the enzyme regulating glucose output in the liver). It induces the expression of proinflammatory cytokines and the suppression of cytokine signaling-3 (a negative regulator of insulin action in the liver). In a study, centrally infusion of resistin excited neuronal activation in the arcuate, paraventricular and dorsomedial nuclei, and it was associated with elevation of neuropeptide Y (NPY) expression in the hypothalamus. Based on this research, resistin can control hepatic insulin sensitivity via NPY neurons [188].

In another study, s-resistin (a non-secretable resistin isoform) can act as an intracrine factor that regulates adipocyte maturation, inflammatory and insulin response in 3T3-L1 cells (murine 3T3-L1-s-res pre-adipocytes). Resistin, via CNS, has shown to induce peripheral and central (hypothalamic) insulin resistance and impairs energy homeostasis [189].

S-resistin is highly expressed in the hypothalamus of Wistar rats and it is a physiological component of the metabolic network that is needed for the regulation of glucose homeostasis. The reduction of central s-resistin expression improves hypothalamic insulin and leptin signaling as well as peripheral insulin sensitivity. So, down-regulation of expression of s-resistin improves insulin sensitivity [190].

The plasma level of resistin increase in obesity and it is associated with the metabolic and cardiovascular disease. It is revealed that increased muscle sympathetic nerve activity (SNA) is a result of obesity, so it can promote the risk of diabetes and cardiovascular disease [191]. In addition, obesity affects baroreceptors and excites renal sympathetic nerve, deregulates adipokine (such as leptin, adiponectin, and resistin) levels, activates the renin-angiotensin system, and increases plasma insulin level. Therefore, obesity can cause renal base hypertension (Figure 5) [192]. The obtained results indicate that resistin elevates renal SNA via PI3K and decreases brown adipose tissue SNA via ERK1/2 [193].

Adipokines act as immunomodulators so, they can contribute to some neuroimmunologic diseases like multiple sclerosis (MS). There is some evidence that shows a link between obesity and MS. Consequently, the role of adipose tissue in the CNS diseases has become more interesting in recent years [194, 195]. Moreover, resistin is also produced in the pituitary gland and by that, there is an important link between neural control of resistin production and lipid metabolism. Elevation of resistin level can cause hypothalamic leptin resistance and increased the risk of atherosclerosis, cardiovascular disease, and insulin resistance [196, 197].

There are also evidences about the participation of adipokines (specially resistin) in some other nervous system diseases including anorexia nervosa and neuromyelitis optica [197, 198]. Like many other hormones, adipokines are under circadian control. Recent data revealed that higher levels of resistin mRNA in the visceral adipose tissuc of mice expressed in late resting phase (Figure 6) [199]. Further research is needed to reveal the circadian patterns of adipokines in human and other animals in the physiological and pathological states.

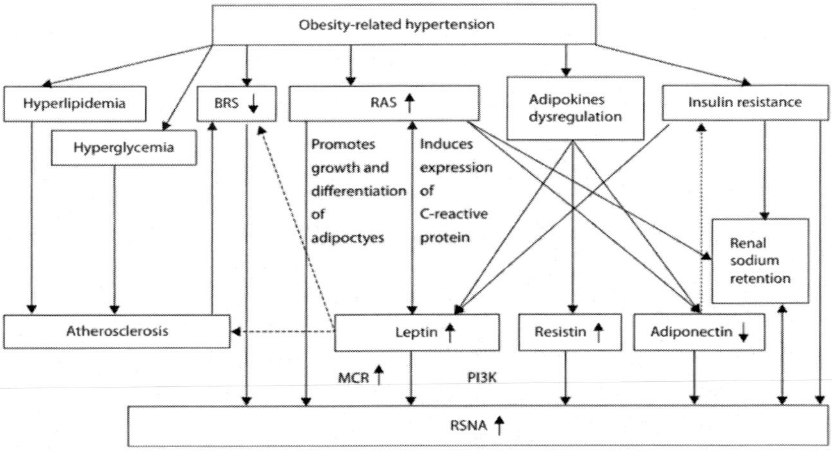


Figure 5. Baroreceptor reflex sensitivity impairs adipokines (such as leptin, adiponectin, and resistin), excessive renin-angiotensin system activation, hyperinsulinemia, and insulin resistance as well as other mechanisms in obesity mutually increase, up-regulate signal pathways, and over activate the renal sympathetic nerve, thus contributes in the development of hypertension. PI3K: phosphatidyl inositol 3 kinase, MCR: melanocortin receptors, RAS: renin-angiotensin system, BRS: baroreceptor reflex sensitivity, RSNA: renal sympathetic nerve activity [192].

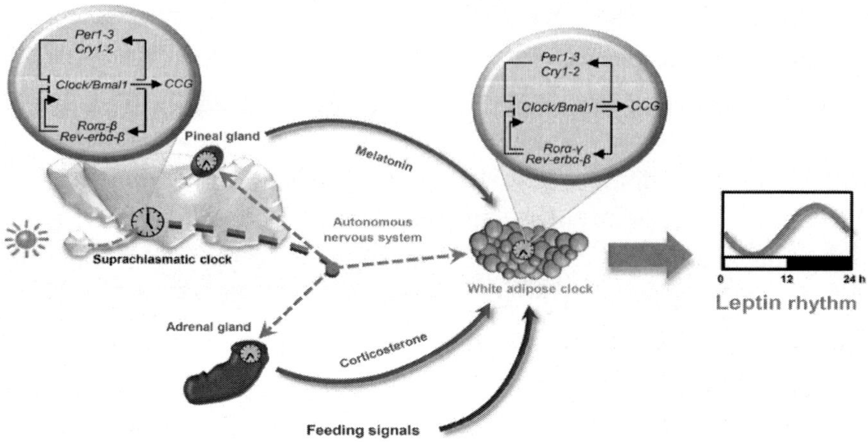


Figure 6. Schematic overview of the control of the circadian rhythm of plasma leptin in nocturnal rodents. The circadian rhythm of leptin secretion is controlled by the local circadian clock in the adipocytes, by temporal cues from the master suprachiasmatic clock via the autonomous nervous system, hormonal rhythms (i.e., pineal melatonin and adrenal glucocorticoids such as corticosterone), and by feeding signals. The insets illustrate a simplified model of the molecular clockwork. Per, period; cry, cryptochromes; CCG, clock-controlled genes [199].

## REFERENCES


[1] Kim KH, Lee K, Moon YS, Sul HS. A cysteine-rich adipose tissue-specific secretory factor inhibits adipocyte differentiation. *Journal of Biological Chemistry*. 2001;276(14):11252-6.

[2] Holcomb IN, Kabakoff RC, Chan B, Baker TW, Gurney A, Henzel W, Nelson C, Lowman HB, Wright BD, Skelton NJ. FIZZ1, a novel cysteine-rich secreted protein associated with pulmonary inflammation, defines a new gene family. *The EMBO journal*. 2000;19(15):4046-55.

[3] Steppan CM, Bailey ST, Bhat S, Brown EJ, Banerjee RR, Wright CM, Patel HR, Ahima RS, Lazar MA. The hormone resistin links obesity to diabetes. *Nature*. 2001;409(6818):307.

[4] Pine GM, Batugedara HM, Nair MG. Here, there and everywhere: Resistin-like molecules in infection, inflammation, and metabolic disorders. *Cytokine.* 2018.

[5] Schinke T, Haberland M, Jamshidi A, Nollau P, Rueger JM, Amling M. Cloning and functional characterization of resistin-like molecule γ. *Biochemical and biophysical research communications.* 2004;314 (2):356-62.

[6] Steppan CM, Brown EJ, Wright CM, Bhat S, Banerjee RR, Dai CY, Enders GH, Silberg DG, Wen X, Wu GD. A family of tissue-specific resistin-like molecules. *Proceedings of the National Academy of Sciences.* 2001;98(2):502-6.

[7] Rajala MW, Lin Y, Ranalletta M, Yang XM, Qian H, Gingerich R, Barzilai N, Scherer PE. Cell type-specific expression and coregulation of murine resistin and resistin-like molecule-α in adipose tissue. *Molecular endocrinology.* 2002;16(8):1920-30.

[8] Nogueiras R, Gallego R, Gualillo O, Caminos JE, García-Caballero T, Casanueva FF, Diéguez C. Resistin is expressed in different rat tissues and is regulated in a tissue-and gender-specific manner. *FEBS letters.* 2003;548(1-3):21-7.

[9] Nogueiras R, Barreiro ML, Caminos JE, Gaytán F, Suominen JS, Navarro VM, Casanueva FF, Aguilar E, Toppari J, Diéguez C. Novel expression of resistin in rat testis: functional role and regulation by nutritional status and hormonal factors. *Journal of cell science.* 2004;117(15):3247-57.

[10] Morash BA, Willkinson D, Ur E, Wilkinson M. Resistin expression and regulation in mouse pituitary. *FEBS letters.* 2002;526(1-3):26-30.

[11] Del Arco A, Peralta S, Carrascosa J, Ros M, Andres A, Arribas C. Alternative splicing generates a novel non-secretable resistin isoform in Wistar rats. *FEBS letters.* 2003;555(2):243-9.

[12] Fain JN, Cheema PS, Bahouth SW, Hiler ML. Resistin release by human adipose tissue explants in primary culture. *Biochemical and biophysical research communications.* 2003;300(3):674-8.

[13] Patel L, Buckels AC, Kinghorn IJ, Murdock PR, Holbrook JD, Plumpton C, Macphee CH, Smith SA. Resistin is expressed in human

macrophages and directly regulated by PPARγ activators. *Biochemical and biophysical research communications.* 2003;300 (2):472-6.

[14] Kang HK, Park J, Seo KS, Kim SH, Choi YJ, Moon YS. Characteristics of structure and expression pattern of ADSF/resistin gene in Korean native cattle. *Asian Australasian Journal of Animal Sciences.* 2006;19(3):329.

[15] Patel SD, Rajala MW, Rossetti L, Scherer PE, Shapiro L. Disulfide-dependent multimeric assembly of resistin family hormones. *Science.* 2004;304(5674):1154-8.

[16] Munir I, Yen HW, Baruth T, Tarkowski R, Azziz R, Magoffin DA, Jakimiuk AJ. Resistin stimulation of 17α-hydroxylase activity in ovarian theca cells *in vitro*: relevance to polycystic ovary syndrome. *The Journal of Clinical Endocrinology & Metabolism.* 2005;90 (8):4852-7.

[17] Rak-Mardyła A, Durak M, Gregoraszczuk EŁ. Effects of resistin on porcine ovarian follicle steroidogenesis in prepubertal animals: an *in vitro* study. *Reproductive Biology and Endocrinology.* 2013;11(1):45.

[18] Gerber M, Boettner A, Seidel B, Lammert A, Bar J, Schuster E, Thiery J, Kiess W, Kratzsch J. Serum resistin levels of obese and lean children and adolescents: biochemical analysis and clinical relevance. *The Journal of Clinical Endocrinology & Metabolism.* 2005;90 (8):4503-9.

[19] Schwartz DR, Lazar MA. Human resistin: found in translation from mouse to man. *Trends in Endocrinology & Metabolism.* 2011;22 (7):259-65.

[20] Park HK, Kwak MK, Kim HJ, Ahima RS. Linking resistin, inflammation, and cardiometabolic diseases. *The Korean journal of internal medicine*. 2017;32(2):239.

[21] Satoh T, Akira S. Toll-Like Receptor Signaling and Its Inducible Proteins. *Microbiology spectrum.* 2016;4(6).

[22] Hussein WM, Liu TY, Skwarczynski M, Toth I. Toll-like receptor agonists: a patent review (2011–2013). *Expert opinion on therapeutic patents.* 2014;24(4):453-70.

[23] Kawasaki T, Kawai T. Toll-like receptor signaling pathways. *Frontiers in immunology*. 2014;5.

[24] Tarkowski A, Bjersing J, Shestakov A, Bokarewa MI. Resistin competes with lipopolysaccharide for binding to toll-like receptor 4. *Journal of cellular and molecular medicine*. 2010;14(6b):1419-31.

[25] Benomar Y, Gertler A, De Lacy P, Crépin D, Hamouda HO, Riffault L, Taouis M. Central resistin overexposure induces insulin resistance through Toll-like receptor 4. *Diabetes*. 2013;62(1):102-14.

[26] Leifer CA, Medvedev AE. Molecular mechanisms of regulation of Toll-like receptor signaling. *Journal of leukocyte biology*. 2016;100(5):927-41.

[27] Brubaker SW, Bonham KS, Zanoni I, Kagan JC. Innate immune pattern recognition: a cell biological perspective. *Annual review of immunology*. 2015;33:257-90.

[28] Jia SJ, Niu PP, Cong JZ, Zhang BK, Zhao M. TLR4 signaling: A potential therapeutic target in ischemic coronary artery disease. *International immunopharmacology*. 2014;23(1):54-9.

[29] Mostafa Mostafazadeh, Sanya Haiaty, Ali Rastqar, Keshvari M. Correlation between Resistin Level and Metabolic Syndrome Component: A Review. *Hormone and Metabolic Research* 2018;50: DOI https://doi.org/10.1055/a-0637-1975.

[30] Reverchon M, Maillard V, Froment P, Ramé C, Dupont J. Adiponectine et résistine-Un rôle dans les fonctions de reproduction? *Médecine/Sciences*. 2013;29(4):417-24. [Adiponectin and resistin-A role in reproductive functions?]

[31] Lee S, Lee HC, Kwon YW, Lee SE, Cho Y, Kim J, Lee S, Kim JY, Lee J, Yang HM. Adenylyl cyclase-associated protein 1 is a receptor for human resistin and mediates inflammatory actions of human monocytes. *Cell metabolism*. 2014;19(3):484-97.

[32] Zhang H, Ghai P, Wu H, Wang C, Field J, Zhou GL. Mammalian adenylyl cyclase-associated protein 1 (CAP1) regulates cofilin function, the actin cytoskeleton, and cell adhesion. *Journal of biological chemistry*. 2013;288(29):20966-77.

[33] Mostafazadeh M, Haiaty S, Rastqar A, Keshvari M. Correlation Between Resistin Level and Metabolic Syndrome Component: A Review. *Hormone and Metabolic Research.* 2018;50(07):521-36.

[34] Jiang S, Park DW, Tadie JM, Gregoire M, Deshane J, Pittet JF, Abraham E, Zmijewski JW. Human resistin promotes neutrophil proinflammatory activation and neutrophil extracellular trap formation and increases severity of acute lung injury. *The Journal of Immunology.* 2014;192(10):4795-803.

[35] Jamaluddin MS, Weakley SM, Yao Q, Chen C. Resistin: functional roles and therapeutic considerations for cardiovascular disease. *British journal of pharmacology.* 2012;165(3):622-32.

[36] Bokarewa M, Nagaev I, Dahlberg L, Smith U, Tarkowski A. Resistin, an adipokine with potent proinflammatory properties. *The Journal of Immunology.* 2005;174(9):5789-95.

[37] Li B, Fang J, Zuo Z, Yin S, He T, Yang M, Deng J, Shen L, Ma X, Yu S. Activation of the porcine alveolar macrophages via toll-like receptor 4/NF-κB mediated pathway provides a mechanism of resistin leading to inflammation. *Cytokine.* 2018.

[38] Zuniga MC, Raghuraman G, Hitchner E, Weyand C, Robinson W, Zhou W. PKC-epsilon and TLR4 synergistically regulate resistin-mediated inflammation in human macrophages. *Atherosclerosis.* 2017;259:51-9.

[39] Jiang CY, Wang W. Resistin aggravates the expression of proinflammatory cytokines in cerulein‑stimulated AR42J pancreatic acinar cells. *Molecular medicine reports.* 2017;15(1):502-6.

[40] Morash BA, Ur E, Wiesner G, Roy J, Wilkinson M. Pituitary resistin gene expression: effects of age, gender and obesity. *Neuro-endocrinology.* 2004;79(3):149-56.

[41] Adeghate E. An update on the biology and physiology of resistin. *Cellular and Molecular Life Sciences CMLS.* 2004;61(19-20): 2485-96.

[42] Brown R, Wiesner G, Ur E, Wilkinson M. Pituitary resistin gene expression is upregulated *in vitro* and *in vivo* by dexamethasone but is unaffected by rosiglitazone. *Neuroendocrinology.* 2005;81(1):41-8.

[43] Rodríguez-Pacheco F, Vázquez-Martínez R, Martínez-Fuentes AJ, Pulido MR, Gahete MD, Vaudry H, Gracia-Navarro F, Dieguez C, Castano JP, Malagon MM. Resistin regulates pituitary somatotrope cell function through the activation of multiple signaling pathways. *Endocrinology.* 2009;150(10):4643-52.

[44] Vázquez MaJ, González CR, Varela L, Lage R, Tovar S, Sangiao-Alvarellos S, Williams LM, Vidal-Puig A, Nogueiras R, López M. Central resistin regulates hypothalamic and peripheral lipid metabolism in a nutritional-dependent fashion. *Endocrinology.* 2008;149(9):4534-43.

[45] Burnett MS, Devaney JM, Adenika RJ, Lindsay R, Howard BV. Cross-sectional associations of resistin, coronary heart disease, and insulin resistance. *The Journal of clinical endocrinology and metabolism.* 2006;91(1):64-8.

[46] Burnett MS, Lee CW, Kinnaird TD, Stabile E, Durrani S, Dullum MK, Devaney JM, Fishman C, Stamou S, Canos D, Zbinden S, Clavijo LC, Jang GJ, Andrews JA, Zhu J, Epstein SE. The potential role of resistin in atherogenesis. *Atherosclerosis.* 2005;182(2):241-8.

[47] Cao H. Adipocytokines in obesity and metabolic disease. *The Journal of endocrinology.* 2014;220(2):T47-59.

[48] Chen BH, Song Y, Ding EL, Roberts CK, Manson JE, Rifai N, Buring JE, Gaziano JM, Liu S. Circulating levels of resistin and risk of type 2 diabetes in men and women: results from two prospective cohorts. *Diabetes care.* 2009;32(2):329-34.

[49] Heidemann C, Sun Q, van Dam RM, Meigs JB, Zhang C, Tworoger SS, Mantzoros CS, Hu FB. Total and high-molecular-weight adiponectin and resistin in relation to the risk for type 2 diabetes in women. *Annals of internal medicine.* 2008;149(5):307-16.

[50] Filkova M, Haluzik M, Gay S, Senolt L. The role of resistin as a regulator of inflammation: Implications for various human pathologies. *Clinical immunology.* 2009;133(2):157-70.

[51] Jamaluddin MS, Weakley SM, Yao Q, Chen C. Resistin: functional roles and therapeutic considerations for cardiovascular disease. *British journal of pharmacology.* 2012;165(3):622-32.

[52] Nogueiras R, Novelle MG, Vazquez MJ, Lopez M, Dieguez C. Resistin: regulation of food intake, glucose homeostasis and lipid metabolism. *Endocrine development.* 2010;17:175-84.

[53] Steppan CM, Bailey ST, Bhat S, Brown EJ, Banerjee RR, Wright CM, Patel HR, Ahima RS, Lazar MA. The hormone resistin links obesity to diabetes. *Nature.* 2001;409:307.

[54] Way JM, Gorgun CZ, Tong Q, Uysal KT, Brown KK, Harrington WW, Oliver WR, Jr., Willson TM, Kliewer SA, Hotamisligil GS. Adipose tissue resistin expression is severely suppressed in obesity and stimulated by peroxisome proliferator-activated receptor gamma agonists. *The Journal of biological chemistry.* 2001;276(28):25651-3.

[55] Fujita H, Fujishima H, Morii T, Koshimura J, Narita T, Kakei M, Ito S. Effect of metformin on adipose tissue resistin expression in db/db mice. *Biochemical and biophysical research communications.* 2002;298(3):345-9.

[56] Bajaj M, Suraamornkul S, Hardies LJ, Pratipanawatr T, DeFronzo RA. Plasma resistin concentration, hepatic fat content, and hepatic and peripheral insulin resistance in pioglitazone-treated type II diabetic patients. *International journal of obesity and related metabolic disorders: Journal of the International Association for the Study of Obesity*. 2004;28(6):783-9.

[57] Savage DB, Sewter CP, Klenk ES, Segal DG, Vidal-Puig A, Considine RV, O'Rahilly S. Resistin / Fizz3 expression in relation to obesity and peroxisome proliferator-activated receptor-gamma action in humans. *Diabetes.* 2001;50(10):2199-202.

[58] Cusi K, Consoli A, DeFronzo RA. Metabolic effects of metformin on glucose and lactate metabolism in noninsulin-dependent diabetes mellitus. *The Journal of clinical endocrinology and metabolism.* 1996;81(11):4059-67.

[59] Smekal A, Vaclavik J. *Adipokines and cardiovascular disease: A comprehensive review.* Biomedical papers of the Medical Faculty of the University Palacky, Olomouc, Czechoslovakia. 2017;161(1): 31-40.

[60] Stumvoll M, Nurjhan N, Perriello G, Dailey G, Gerich JE. Metabolic effects of metformin in non-insulin-dependent diabetes mellitus. *The New England journal of medicine.* 1995;333(9):550-4.

[61] Wiernsperger NF, Bailey CJ. The antihyperglycaemic effect of metformin: therapeutic and cellular mechanisms. *Drugs.* 1999;58 Suppl 1:31-9; discussion 75-82.

[62] Reilly MP, Lehrke M, Wolfe ML, Rohatgi A, Lazar MA, Rader DJ. Resistin is an inflammatory marker of atherosclerosis in humans. *Circulation.* 2005;111(7):932-9.

[63] Arnaud C, Braunersreuther V, Mach F. Toward immunomodulatory and anti-inflammatory properties of statins. *Trends in cardiovascular medicine.* 2005;15(6):202-6.

[64] Asgary S, SamsamShariat SZ, Ghorbani A, Keshvari M, Sahebkar A, Sarrafzadegan N. Relationship between serum resistin concentrations with metabolic syndrome and its components in an Iranian population. *Diabetes & Metabolic Syndrome: Clinical Research & Reviews.* 2015;9(4):266-70.

[65] Hu WL, Qiao SB, Li JJ. Decreased C-reactive protein-induced resistin production in human monocytes by simvastatin. *Cytokine.* 2007;40(3):201-6.

[66] Ichida Y, Hasegawa G, Fukui M, Obayashi H, Ohta M, Fujinami A, Ohta K, Nakano K, Yoshikawa T, Nakamura N. Effect of atorvastatin on *in vitro* expression of resistin in adipocytes and monocytes/ macrophages and effect of atorvastatin treatment on serum resistin levels in patients with type 2 diabetes. *Pharmacology.* 2006;76(1):34-9.

[67] Shyu KG, Chua SK, Wang BW, Kuan P. Mechanism of inhibitory effect of atorvastatin on resistin expression induced by tumor necrosis factor-alpha in macrophages. *Journal of biomedical science.* 2009; 16:50.

[68] Silswal N, Singh AK, Aruna B, Mukhopadhyay S, Ghosh S, Ehtesham NZ. Human resistin stimulates the pro-inflammatory cytokines TNF-alpha and IL-12 in macrophages by NF-kappa B-dependent pathway.

*Biochemical and biophysical research communications*. 2005;334 (4):1092-101.

[69] Schaffler A, Ehling A, Neumann E, Herfarth H, Tarner I, Scholmerich J, Muller-Ladner U, Gay S. Adipocytokines in synovial fluid. *JAMA*. 2003;290(13):1709-10.

[70] Migita K, Maeda Y, Miyashita T, Kimura H, Nakamura M, Ishibashi H, Eguchi K. The serum levels of resistin in rheumatoid arthritis patients. *Clinical and experimental rheumatology*. 2006;24(6): 698-701.

[71] Senolt L, Housa D, Vernerova Z, Jirasek T, Svobodova R, Veigl D, Anderlova K, Muller-Ladner U, Pavelka K, Haluzik M. Resistin in rheumatoid arthritis synovial tissue, synovial fluid and serum. *Annals of the rheumatic diseases*. 2007;66(4):458-63.

[72] Imperato AK, Bingham CO, 3rd, Abramson SB. Overview of benefit/ risk of biological agents. *Clinical and experimental rheumatology*. 2004;22(5 Suppl 35):S108-14.

[73] Maini RN, Breedveld FC, Kalden JR, Smolen JS, Davis D, Macfarlane JD, Antoni C, Leeb B, Elliott MJ, Woody JN, Schaible TF, Feldmann M. Therapeutic efficacy of multiple intravenous infusions of anti-tumor necrosis factor alpha monoclonal antibody combined with low-dose weekly methotrexate in rheumatoid arthritis. *Arthritis and rheumatism*. 1998;41(9):1552-63.

[74] Gonzalez-Gay MA, Garcia-Unzueta MT, Gonzalez-Juanatey C, Miranda-Filloy JA, Vazquez-Rodriguez TR, De Matias JM, Martin J, Dessein PH, Llorca J. Anti-TNF-alpha therapy modulates resistin in patients with rheumatoid arthritis. *Clinical and experimental rheumatology*. 2008;26(2):311-6.

[75] Koh KK, Quon MJ, Lee SJ, Han SH, Ahn JY, Kim JA, Chung WJ, Lee Y, Shin EK. Efonidipine simultaneously improves blood pressure, endothelial function, and metabolic parameters in nondiabetic patients with hypertension. *Diabetes care*. 2007;30(6):1605-7.

[76] Westphal S, Borucki K, Taneva E, Makarova R, Luley C. Adipokines and treatment with niacin. *Metabolism: clinical and experimental*. 2006;55(10):1283-5.

[77] Bo S, Ciccone G, Durazzo M, Gambino R, Massarenti P, Baldi I, Lezo A, Tiozzo E, Pauletto D, Cassader M, Pagano G. Efficacy of antioxidant treatment in reducing resistin serum levels: a randomized study. *PLoS clinical trials.* 2007;2(5):e17.

[78] Heilbronn LK, Rood J, Janderova L, Albu JB, Kelley DE, Ravussin E, Smith SR. Relationship between serum resistin concentrations and insulin resistance in nonobese, obese, and obese diabetic subjects. *The Journal of clinical endocrinology and metabolism.* 2004;89(4): 1844-8.

[79] Holcomb IN, Kabakoff RC, Chan B, Baker TW, Gurney A, Henzel W, Nelson C, Lowman HB, Wright BD, Skelton NJ, Frantz GD, Tumas DB, Peale FV, Jr., Shelton DL, Hebert CC. FIZZ1, a novel cysteine-rich secreted protein associated with pulmonary inflammation, defines a new gene family. *The EMBO journal.* 2000;19(15):4046-55.

[80] Reinehr T, Roth CL, Menke T, Andler W. Resistin concentrations before and after weight loss in obese children. *International journal of obesity.* 2006;30(2):297-301.

[81] Utzschneider KM, Carr DB, Tong J, Wallace TM, Hull RL, Zraika S, Xiao Q, Mistry JS, Retzlaff BM, Knopp RH, Kahn SE. Resistin is not associated with insulin sensitivity or the metabolic syndrome in humans. *Diabetologia.* 2005;48(11):2330-3.

[82] Vozarova de Courten B, Degawa-Yamauchi M, Considine RV, Tataranni PA. High serum resistin is associated with an increase in adiposity but not a worsening of insulin resistance in Pima Indians. *Diabetes.* 2004;53(5):1279-84.

[83] Seto SW, Lam TY, Or PM, Lee WY, Au AL, Poon CC, Li RW, Chan SW, Yeung JH, Leung GP, Lee SM, Ngai SM, Kwan YW. Folic acid consumption reduces resistin level and restores blunted acetylcholine-induced aortic relaxation in obese/diabetic mice. *The Journal of nutritional biochemistry.* 2010;21(9):872-80.

[84] Felipe F, Bonet ML, Ribot J, Palou A. Modulation of resistin expression by retinoic acid and vitamin A status. *Diabetes.* 2004;53(4):882-9.

[85] Kougias P, Chai H, Lin PH, Lumsden AB, Yao Q, Chen C. Adipocyte-derived cytokine resistin causes endothelial dysfunction of porcine coronary arteries. *Journal of vascular surgery.* 2005;41(4):691-8.

[86] Haugen F, Jørgensen A, Drevon CA, Trayhurn P. Inhibition by insulin of resistin gene expression in 3T3-L1 adipocytes. *FEBS letters.* 2001;507(1):105-8.

[87] Asensio Cd, Cettour-Rose P, Theander-Carrillo C, Rohner-Jeanrenaud Fo, Muzzin P. Changes in glycemia by leptin administration or high-fat feeding in rodent models of obesity/type 2 diabetes suggest a link between resistin expression and control of glucose homeostasis. *Endocrinology.* 2004;145(5):2206-13.

[88] Sánchez-Solana B, Laborda J, Baladrón V. Mouse resistin modulates adipogenesis and glucose uptake in 3T3-L1 preadipocytes through the ROR1 receptor. *Molecular endocrinology.* 2012;26(1):110-27.

[89] Ikeda Y, Tsuchiya H, Hama S, Kajimoto K, Kogure K. Resistin affects lipid metabolism during adipocyte maturation of 3T3-L1 cells. *The FEBS journal.* 2013;280(22):5884-95.

[90] He F, Jin JQ, Qin QQ, Zheng YQ, Li TT, Zhang Y, He JD. Resistin Regulates Fatty Acid Β Oxidation by Suppressing Expression of Peroxisome Proliferator Activator Receptor Gamma-Coactivator 1α (PGC-1α). *Cellular Physiology and Biochemistry.* 2018;46(5): 2165-72.

[91] Kim KH, Zhao L, Moon Y, Kang C, Sul HS. Dominant inhibitory adipocyte-specific secretory factor (ADSF)/resistin enhances adipogenesis and improves insulin sensitivity. *Proceedings of the National Academy of Sciences*. 2004;101(17):6780-5.

[92] Brunetti L, Orlando G, Recinella L, Michelotto B, Ferrante C, Vacca M. Resistin, but not adiponectin, inhibits dopamine and norepinephrine release in the hypothalamus. *European journal of pharmacology*. 2004;493(1-3):41-4.

[93] Lazar M. Resistin-and obesity-associated metabolic diseases. *Hormone and Metabolic Research.* 2007;39(10):710-6.

[94] Tovar S, Nogueiras R, Tung LY, Castaneda TR, Vázquez MJ, Morris A, Williams LM, Dickson SL, Diéguez C. Central administration of

resistin promotes short-term satiety in rats. *European Journal of Endocrinology*. 2005;153(3):R1-R5.

[95] Muse ED, Obici S, Bhanot S, Monia BP, McKay RA, Rajala MW, Scherer PE, Rossetti L. Role of resistin in diet-induced hepatic insulin resistance. *The Journal of clinical investigation*. 2004;114(2):232-9.

[96] Lee JH, Bullen Jr JW, Stoyneva VL, Mantzoros CS. Circulating resistin in lean, obese, and insulin-resistant mouse models: lack of association with insulinemia and glycemia. *American Journal of Physiology-Endocrinology and Metabolism*. 2005;288(3):E625-E32.

[97] Oliver P, Ribot J, Rodriguez A, Sanchez J, Pico C, Palou A. Resistin as a putative modulator of insulin action in the daily feeding/fasting rhythm. *Pflügers Archiv*. 2006;452(3):260.

[98] Schaalan M, El-Abhar HS, Barakat M, El-Denshary ES. Westernized-like-diet-fed rats: effect on glucose homeostasis, lipid profile, and adipocyte hormones and their modulation by rosiglitazone and glimepiride. *Journal of diabetes and its complications*. 2009;23 (3):199-208.

[99] Oliveira MC, Menezes-Garcia Z, Henriques MC, Soriani FM, Pinho V, Faria AM, Santiago AF, Cara DC, Souza DG, Teixeira MM. Acute and sustained inflammation and metabolic dysfunction induced by high refined carbohydrate-containing diet in mice. *Obesity*. 2013;21(9):E396-E406.

[100] Barbosa-da-Silva S, Fraulob-Aquino JC, Lopes JR, Mandarim-de-Lacerda CA, Aguila MB. Weight cycling enhances adipose tissue inflammatory responses in male mice. *PloS one*. 2012;7(7):e39837.

[101] Juan CC, Au LC, Fang VS, Kang SF, Ko YH, Kuo SF, Hsu YP, Kwok CF, Ho LT. Suppressed gene expression of adipocyte resistin in an insulin-resistant rat model probably by elevated free fatty acids. *Biochemical and biophysical research communications*. 2001;289 (5):1328-33.

[102] Le Lay S, Boucher J, Rey A, Castan-Laurell I, Krief S, Ferré P, Valet P, Dugail I. Decreased resistin expression in mice with different sensitivities to a high-fat dieot. *Biochemical and biophysical research communications*. 2001;289(2):564-7.

[103] Masuzaki H, Paterson J, Shinyama H, Morton NM, Mullins JJ, Seckl JR, Flier JS. A transgenic model of visceral obesity and the metabolic syndrome. *Science.* 2001;294(5549):2166-70.
[104] Way JM, Görgün CZ, Tong Q, Uysal KT, Brown KK, Harrington WW, Oliver WR, Willson TM, Kliewer SA, Hotamisligil GS. Adipose tissue resistin expression is severely suppressed in obesity and stimulated by peroxisome proliferator-activated receptor γ agonists. *Journal of Biological Chemistry.* 2001;276(28):25651-3.
[105] Kusminski CM, Mcternan PG, Kumar S. Role of resistin in obesity, insulin resistance and Type II diabetes. *Clinical science.* 2005;109(3):243-56.
[106] Graveleau C, Zaha VG, Mohajer A, Banerjee RR, Dudley-Rucker N, Steppan CM, Rajala MW, Scherer PE, Ahima RS, Lazar MA. Mouse and human resistin impair glucose transport in primary mouse cardiomyocytes and oligomerization is required for this biological action. *Journal of biological chemistry.* 2005.
[107] Palanivel R, Maida A, Liu Y, Sweeney G. Regulation of insulin signalling, glucose uptake and metabolism in rat skeletal muscle cells upon prolonged exposure to resistin. *Diabetologia.* 2006;49(1): 183-90.
[108] Rajala MW, Obici S, Scherer PE, Rossetti L. Adipose-derived resistin and gut-derived resistin-like molecule–β selectively impair insulin action on glucose production. *The Journal of clinical investigation.* 2003;111(2):225-30.
[109] Sassek M, Pruszynska-Oszmalek E, Kołodziejski PA, Szczepankiewicz D, Kaczmarek P, Wieloch M, Kurto K, Nogowski L, Nowak KW, Strowski MZ. Resistin is produced by rat pancreatic islets and regulates insulin and glucagon *in vitro* secretion. *Islets.* 2016; 8(6):177-85.
[110] Song R, Wang X, Mao Y, Li H, Li Z, Xu W, Wang R, Guo T, Jin L, Zhang X. Resistin disrupts glycogen synthesis under high insulin and high glucose levels by down-regulating the hepatic levels of GSK3β. *Gene.* 2013;529(1):50-6.

[111] Banerjee RR, Rangwala SM, Shapiro JS, Rich AS, Rhoades B, Qi Y, Wang J, Rajala MW, Pocai A, Scherer PE. Regulation of fasted blood glucose by resistin. *Science.* 2004;303(5661):1195-8.

[112] Daquinag AC ZY, Amaya-Manzanares F, Simmons PJ, Kolonin MG. An isoform of decorin is a resistin receptor on the surface of adipose progenitor cells. *Cell Stem Cell.* 2011;9:74-86.

[113] Sato N, Kobayashi K, Inoguchi T, Sonoda N, Imamura M, Sekiguchi N, Nakashima N, Nawata H. Adenovirus-mediated high expression of resistin causes dyslipidemia in mice. *Endocrinology.* 2005;146 (1):273-9.

[114] Steppan CM, Wang J, Whiteman EL, Birnbaum MJ, Lazar MA. Activation of SOCS-3 by resistin. *Molecular and cellular biology.* 2005;25(4):1569-75.

[115] Park HK, Ahima RS. Resistin in rodents and humans. *Diabetes & metabolism journal.* 2013;37(6):404-14.

[116] McTernan CL, McTernan P, Harte A, Levick P, Barnett A, Kumar S. Resistin, central obesity, and type 2 diabetes. *The Lancet.* 2002; 359(9300):46-7.

[117] Nagaev I, Smith U. Insulin resistance and type 2 diabetes are not related to resistin expression in human fat cells or skeletal muscle. *Biochemical and biophysical research communications.* 2001;285 (2):561-4.

[118] Savage DB, Sewter CP, Klenk ES, Segal DG, Vidal-Puig A, Considine RV, O'Rahilly S. Resistin/Fizz3 expression in relation to obesity and peroxisome proliferator–activated receptor-γ action in humans. *Diabetes.* 2001;50(10):2199-202.

[119] Tomaru T, Steger DJ, Lefterova MI, Schupp M, Lazar MA. Adipocyte-specific expression of murine resistin is mediated by synergism between peroxisome proliferator-activated receptor γ and CCAAT/enhancer binding proteins. *Journal of Biological Chemistry.* 2009.

[120] Qatanani M, Szwergold NR, Greaves DR, Ahima RS, Lazar MA. Macrophage-derived human resistin exacerbates adipose tissue

inflammation and insulin resistance in mice. *The Journal of clinical investigation.* 2009;119(3):531-9.

[121] Zhang J, Qin Y, Zheng X, Qiu J, Gong L, Mao H, Jia W, Guo J. The relationship between human serum resistin level and body fat content, plasma glucose as well as blood pressure. *Zhonghua yi xue za zhi.* 2002;82(23):1609-12.

[122] Yannakoulia M, Yiannakouris N, Blüher S, Matalas A-L, Klimis-Zacas D, Mantzoros CS. Body fat mass and macronutrient intake in relation to circulating soluble leptin receptor, free leptin index, adiponectin, and resistin concentrations in healthy humans. *The Journal of Clinical Endocrinology & Metabolism.* 2003;88(4):1730-6.

[123] Azuma K, Katsukawa F, Oguchi S, Murata M, Yamazaki H, Shimada A, Saruta T. Correlation between serum resistin level and adiposity in obese individuals. *Obesity research.* 2003;11(8):997-1001.

[124] Jain SH, Massaro JM, Hoffmann U, Rosito GA, Vasan RS, Raji A, O'donnell CJ, Meigs JB, Fox CS. Cross-sectional associations between abdominal and thoracic adipose tissue compartments and adiponectin and resistin in the Framingham Heart Study. *Diabetes* care. 2009.

[125] Caparrós EP, Gómez FI, Ortega MG, Soler IO, Paredes MP, Sampedro EA, Tafalla MA. Resistin in morbidly obese patients before and after gastric bypass surgery. *Nutricion hospitalaria.* 2017; 34(6):1333-7.

[126] Amato MC, Pizzolanti G, Torregrossa V, Misiano G, Milano S, Giordano C. Visceral adiposity index (VAI) is predictive of an altered adipokine profile in patients with type 2 diabetes. *PLoS One.* 2014;9(3):e91969.

[127] De Luis D, Terroba M, Cuellar L, Conde R, Primo D, Aller R, Sagrado M, Izaola O. Resistin levels in morbid obese patients following the biliopancreatic diversion surgery. *Hormone and Metabolic Research.* 2011;43(03):205-8.

[128] Moschen AR, Molnar C, Wolf AM, Weiss H, Graziadei I, Kaser S, Ebenbichler CF, Stadlmann S, Moser PL, Tilg H. Effects of weight

loss induced by bariatric surgery on hepatic adipocytokine expression. *Journal of hepatology.* 2009;51(4):765-77.

[129] Lee JH, Chan JL, Yiannakouris N, Kontogianni M, Estrada E, Seip R, Orlova C, Mantzoros CS. Circulating resistin levels are not associated with obesity or insulin resistance in humans and are not regulated by fasting or leptin administration: cross-sectional and interventional studies in normal, insulin-resistant, and diabetic subjects. *The Journal of Clinical Endocrinology & Metabolism.* 2003;88(10):4848-56.

[130] Wolfe BE, Jimerson DC, Orlova C, Mantzoros CS. Effect of dieting on plasma leptin, soluble leptin receptor, adiponectin and resistin levels in healthy volunteers. *Clinical endocrinology.* 2004;61(3): 332-8.

[131] Hivert M-F, Sullivan LM, Fox CS, Nathan DM, D'Agostino Sr RB, Wilson PW, Meigs JB. Associations of adiponectin, resistin, and tumor necrosis factor-α with insulin resistance. *The Journal of Clinical Endocrinology & Metabolism.* 2008;93(8):3165-72.

[132] Menzaghi C, Coco A, Salvemini L, Thompson R, De Cosmo S, Doria A, Trischitta V. Heritability of serum resistin and its genetic correlation with insulin resistance-related features in nondiabetic Caucasians. *The Journal of Clinical Endocrinology & Metabolism.* 2006;91(7):2792-5.

[133] Osawa H, Tabara Y, Kawamoto R, Ohashi J, Ochi M, Onuma H, Nishida W, Yamada K, Nakura J, Kohara K. Plasma resistin, associated with single nucleotide polymorphism− 420, is correlated with insulin resistance, lower HDL cholesterol, and high-sensitivity C-reactive protein in the Japanese general population. *Diabetes care.* 2007;30(6):1501-6.

[134] Heilbronn L, Rood J, Janderova L, Albu J, Kelley D, Ravussin E, Smith S. Relationship between serum resistin concentrations and insulin resistance in nonobese, obese, and obese diabetic subjects. *The Journal of Clinical Endocrinology & Metabolism.* 2004;89(4):1844-8.

[135] Chen YH, Lee MJ, Chang HH, Hung PF, Kao YH. 17β-Estradiol stimulates resistin gene expression in 3T3-L1 adipocytes via the

estrogen receptor, extracellularly regulated kinase, and CCAAT/ enhancer binding protein-α pathways. *Endocrinology.* 2006;147 (9):4496-504.

[136] Youn BS, Yu KY, Park HJ, Lee NS, Min SS, Youn MY, Cho YM, Park YJ, Kim SY, Lee HK. Plasma resistin concentrations measured by enzyme-linked immunosorbent assay using a newly developed monoclonal antibody are elevated in individuals with type 2 diabetes mellitus. *The Journal of Clinical Endocrinology & Metabolism.* 2004; 89(1):150-6.

[137] Cherneva RV, Georgiev OB, Petrova DS, Mondeshki TL, Ruseva SR, Cakova AD, VI. M. Resistin-the link between adipose tissue dysfunction and insulin resistance in patients with obstructive sleep apnea. *Journal of Diabetes & Metabolic Disorders.* 2013;1(5).

[138] Lobo TF, Torloni MR, Gueuvoghlanian-Silva BY, Mattar R DS. Resistin concentration and gestational diabetes: a systematic review of the literature. *Journal of reproductive immunology.* 2013;97 (1):120-7.

[139] Cebeci E, Cakan C, Gursu M, Uzun S, Karadag S, Koldas M, Calhan T, Helvaci SA, Ozturk S. The Main Determinants of Serum Resistin Level in Type 2 Diabetic Patients are Renal Function and Inflammation not Presence of Microvascular Complication, Obesity and Insulin Resistance. *Experimental and Clinical Endocrinology & Diabetes.* 2018.

[140] Niu XH, Li L, Li JY, Song Q, Jin MM, Liu JX. Serum resistin positively correlates with serum lipids, but not with insulin resistance, in first-degree relatives of type-2 diabetes patients: an observational study in China. *Medicine.* 2017;96(16).

[141] Johnson AM, JM. O. The origins and drivers of insulin resistance. *Cell.* 2013;152(4):673-84.

[142] Kaser S KA, Sandhofer A, Ebenbichler CF, Tilg H, Patsch JR. Resistin messenger-RNA expression is increased by proinflammatory cytokines *in vitro. Biochemical and biophysical research communications.* 2003;309(2):286-90.

[143] Reilly MP LM, Wolfe ML, Rohatgi A, Lazar MA, Rader DJ. Resistin is an inflammatory marker of atherosclerosis in humans. *Circulation.* 2005;111(7):932-9.

[144] Shetty GK EP, Horton ES, Mantzoros CS, Veves A. Circulating adiponectin and resistin levels in relation to metabolic factors, inflammatory markers, and vascular reactivity in diabetic patients and subjects at risk for diabetes. *Diabetes care.* 2004;27(10):2450-7.

[145] Abate N SSH, Rizzo M, Nikolic D, Obradovic M, Bjelogrlic P, R Isenovic E. Resistin: an inflammatory cytokine. Role in cardiovascular diseases, diabetes and the metabolic syndrome. *Current pharmaceutical design.* 2014;20(31):4961-9.

[146] Al Hannan F CK. Human resistin and the RELM of Inflammation in diabesity. *Diabetology & metabolic syndrome.* 2015;7(1):54.

[147] Zhang Y, Li Y, Yu L, Zhou L. Association between serum resistin concentration and hypertension: A systematic review and meta-analysis. *Oncotarget.* 2017;8(25):41529.

[148] Hsu WY, Chao YW, Tsai YL, Lien CC, Chang CF, Deng MC, Ho LT, Kwok CF, Juan CC. Resistin induces monocyte–endothelial cell adhesion by increasing ICAM-1 and VCAM-1 expression in endothelial cells via p38MAPK-dependent pathway. *Journal of cellular physiology.* 2011;226(8):2181-8.

[149] Manduteanu I, Pirvulescu M, Gan AM, Stan D, Simion V, Dragomir E, Calin M, Manea A, Simionescu M. Similar effects of resistin and high glucose on P-selectin and fractalkine expression and monocyte adhesion in human endothelial cells. *Biochemical and biophysical research communications.* 2010;391(3):1443-8.

[150] Ouchi N, Parker JL, Lugus JJ, Walsh K. Adipokines in inflammation and metabolic disease. *Nature reviews immunology.* 2011;11(2):85.

[151] Jamaluddin MS, Yan S, Lü J, Liang Z, Yao Q, Chen C. Resistin increases monolayer permeability of human coronary artery endothelial cells. *PLoS One.* 2013;8(12):e84576.

[152] Chen C, Jiang J, Lü JM, Chai H, Wang X, Lin PH, Yao Q. Resistin decreases expression of endothelial nitric oxide synthase through oxidative stress in human coronary artery endothelial cells. *American*

*Journal of Physiology-Heart and Circulatory Physiology.* 2010;299 (1):H193-H201.

[153] Frühbeck G. *Vasoactive factors and inflammatory mediators produced in adipose tissue. Adipose tissue and adipokines in health and disease:* Springer; 2007. p. 63-77.

[154] Fantuzzi G, Braunschweig C. *Adipose tissue and adipokines in health and disease:* Springer; 2014.

[155] Calabro P, Samudio I, Willerson JT, Yeh ET. Resistin promotes smooth muscle cell proliferation through activation of extracellular signal–regulated kinase 1/2 and phosphatidylinositol 3-kinase pathways. *Circulation.* 2004;110(21):3335-40.

[156] Jiang C, Zhang H, Zhang W, Kong W, Zhu Y, Zhang H, Xu Q, Li Y, Wang X. Homocysteine promotes vascular smooth muscle cell migration by induction of the adipokine resistin. *American Journal of Physiology-Cell Physiology.* 2009;297(6):C1466-C76.

[157] Codoñer-Franch P, Tavárez-Alonso S, Porcar-Almela M, Navarro-Solera M, Arilla-Codoñer Á, Alonso-Iglesias E. Plasma resistin levels are associated with homocysteine, endothelial activation, and nitrosative stress in obese youths. *Clinical biochemistry.* 2014;47(1-2):44-8.

[158] Qiu W, Chen N, Zhang Q, Zhuo L, Wang X, Wang D, Jin H. Resistin increases platelet P-selectin levels via p38 MAPK signal pathway. *Diabetes and Vascular Disease Research.* 2014;11(2):121-4.

[159] Joshi PL. *Effects of resistin on platelet function and its receptor, adenylyl cyclase associated protein 1:* University of York; 2017.

[160] Melone M, Wilsie L, Palyha O, Strack A, Rashid S. Discovery of a new role of human resistin in hepatocyte low-density lipoprotein receptor suppression mediated in part by proprotein convertase subtilisin/kexin type 9. *Journal of the American College of Cardiology.* 2012;59(19):1697-705.

[161] Costandi J, Melone M, Zhao A, Rashid S. Human resistin stimulates hepatic overproduction of atherogenic ApoB-containing lipoprotein particles by enhancing ApoB stability and impairing intracellular

insulin signaling. *Circulation research.* 2011:CIRCRESAHA. 110. 238949.

[162] Lee TS, Lin CY, Tsai JY, Wu YL, Su KH, Lu KY, Hsiao SH, Pan C-C, Kou YR, Hsu Y-P. Resistin increases lipid accumulation by affecting class A scavenger receptor, CD36 and ATP-binding cassette transporter-A1 in macrophages. *Life sciences.* 2009;84(3-4):97-104.

[163] Balistreri CR, Caruso C, Candore G. The role of adipose tissue and adipokines in obesity-related inflammatory diseases. *Mediators of inflammation.* 2010;2010.

[164] Anfossi G, Russo I, Doronzo G, Pomero A, Trovati M. Adipocytokines in atherothrombosis: focus on platelets and vascular smooth muscle cells. *Mediators of inflammation.* 2010;2010.

[165] Ding Q, White SP, Ling C, Zhou W. Resistin and cardiovascular disease. *Trends in cardiovascular medicine.* 2011;21(1):20-7.

[166] Khera AV, Qamar A, Murphy SA, Cannon CP, Sabatine MS, Rader DJ. On-statin resistin, leptin, and risk of recurrent coronary events after hospitalization for an acute coronary syndrome (from the pravastatin or atorvastatin evaluation and infection therapy-thrombolysis in myocardial infarction 22 study). *The American journal of cardiology.* 2015;116(5):694-8.

[167] Huang X, Yang Z. Resistin's, obesity and insulin resistance: the continuing disconnect between rodents and humans. *Journal of endocrinological investigation.* 2016;39(6):607-15.

[168] Piestrzeniewicz K, Luczak K, Goch JH. Value of blood adipose tissue hormones concentration--adiponectin, resistin and leptin in the prediction of major adverse cardiac events (MACE) in 1-year follow-up after primary percutaneous coronary intervention in ST-segment elevation acute myocardial infarction. *Neuro endocrinology letters.* 2008;29(4):581-8.

[169] Hoefle G, Saely CH, Risch L, Koch L, Schmid F, Rein P, Aczél S, Berchtold S, Drexel H. Relationship between the adipose-tissue hormone resistin and coronary artery disease. *Clinica chimica acta.* 2007;386(1-2):1-6.

[170] Muse ED, Feldman DI, Blaha MJ, Dardari ZA, Blumenthal RS, Budoff MJ, Nasir K, Criqui MH, Cushman M, McClelland RL. The association of resistin with cardiovascular disease in the Multi-Ethnic Study of Atherosclerosis. *Atherosclerosis.* 2015;239(1):101-8.

[171] Mu H, Ohashi R, Yan S, Chai H, Yang H, Lin P, Yao Q, Chen C. Adipokine resistin promotes *in vitro* angiogenesis of human endothelial cells. *Cardiovascular research.* 2006;70(1):146-57.

[172] Di Simone N, Di Nicuolo F, Sanguinetti M, Castellani R, D'Asta M, Caforio L, Caruso A. Resistin regulates human choriocarcinoma cell invasive behaviour and endothelial cell angiogenic processes. *Journal of endocrinology.* 2006;189(3):691-9.

[173] Li YM, Zhou BP, Deng J, Pan Y, Hay N, Hung M-C. A hypoxia-independent hypoxia-inducible factor-1 activation pathway induced by phosphatidylinositol-3 kinase/Akt in HER2 overexpressing cells. *Cancer Research.* 2005;65(8):3257-63.

[174] Fang W, Zhang Q, Peng Y, Chen M, Lin X, Wu J, Cai C, Mei Y, Jin H. Resistin levelispositively correlated with thrombotic complications in Southern Chinese metabolic syndrome patients. *Journal of endocrinological investigation.* 2011;34(2):e36-e42.

[175] Moore C, Sanz-Rosa D, Emerson M. Distinct role and location of the endothelial isoform of nitric oxide synthase in regulating platelet aggregation in males and females *in vivo*. *European journal of pharmacology.* 2011;651(1-3):152-8.

[176] Takata Y, Osawa H, Kurata M, Kurokawa M, Yamauchi J, Ochi M, Nishida W, Okura T, Higaki J, Makino H. Hyperresistinemia is associated with coexistence of hypertension and type 2 diabetes. *Hypertension.* 2008;51(2):534-9.

[177] Thomopoulos C, Daskalaki M, Papazachou O, Rodolakis N, Bratsas A, Papadopoulos D, Papavasileiou M, Perrea D, Makris T. Association of resistin and adiponectin with different clinical blood pressure phenotypes. *Journal of human hypertension.* 2011;25(1):38.

[178] Papadopoulos D, Makris TK, Perrea D, Papazachou O, Daskalaki M, Sanidas E, Votteas V. Adiponectin–insulin and resistin plasma levels

in young healthy offspring of patients with essential hypertension. *Blood pressure.* 2008;17(1):50-4.

[179] Kaser S, Kaser A, Sandhofer A, Ebenbichler C, Tilg H, Patsch J. Resistin messenger-RNA expression is increased by proinflammatory cytokines *in vitro*. *Biochemical and biophysical research communications.* 2003;309(2):286-90.

[180] Hu WL, Qian SB, Li JJ. Decreased C-reactive protein-induced resistin production in human monocytes by simvastatin. *Cytokine.* 2007;40(3):201-6.

[181] Nagaev I, Bokarewa M, Tarkowski A, Smith U. Human resistin is a systemic immune-derived proinflammatory cytokine targeting both leukocytes and adipocytes. *PloS one.* 2006;1(1):e31.

[182] Lu DY, Chen JH, Tan TW, Huang CY, Yeh WL, Hsu HC. Resistin protects against 6-hydroxydopamine-induced cell death in dopaminergic-like MES23. 5 cells. *Journal of cellular physiology.* 2013;228(3):563-71.

[183] Berghoff M, Hochberg A, Schmid A, Schlegel J, Karrasch T, Kaps M, Schäffler A. Quantification and regulation of the adipokines resistin and progranulin in human cerebrospinal fluid. *European journal of clinical investigation.* 2016;46(1):15-26.

[184] Wang CH, Wang WT, Cheng SY, Hung WT, Wu TL, Hsueh C-M. Leptin and Interleukin-1β Modulate Neuronal Glutamate Release and Protect Against Glucose-Oxygen-Serum Deprivation. *Current neurovascular research.* 2010;7(3):223-37.

[185] Perez-Gonzalez R, Antequera D, Vargas T, Spuch C, Bolós M, Carro E. Leptin induces proliferation of neuronal progenitors and neuroprotection in a mouse model of Alzheimer's disease. *Journal of Alzheimer's Disease.* 2011;24(supplement 2):17-25.

[186] Qiu G, Wan R, Hu J, Mattson MP, Spangler E, Liu S, Yau S-Y, Lee TM, Gleichmann M, Ingram DK. Adiponectin protects rat hippocampal neurons against excitotoxicity. *Age.* 2011;33(2):155-65.

[187] Parimisetty A, Dorsemans AC, Awada R, Ravanan P, Diotel N, d'Hellencourt CL. Secret talk between adipose tissue and central nervous system via secreted factors—an emerging frontier in the

neurodegenerative research. *Journal of neuroinflammation.* 2016;13 (1):67.

[188] Singhal NS, Lazar MA, Ahima RS. Central resistin induces hepatic insulin resistance via neuropeptide Y. *Journal of Neuroscience.* 2007;27(47):12924-32.

[189] Inoue H. Central insulin-mediated regulation of hepatic glucose production. *Endocrine journal.* 2016;63(1):1-7.

[190] Rodríguez M, Pintado C, Moltó E, Gallardo N, Fernández-Martos CM, López V, Andrés A, Arribas C. Central s-resistin deficiency ameliorates hypothalamic inflammation and increases whole body insulin sensitivity. *Scientific reports.* 2018;8(1):3921.

[191] Kosari S, Rathner J, Chen F, Kosari S, Badoer E. Centrally administered resistin enhances sympathetic nerve activity to the hindlimb but attenuates the activity to brown adipose tissue. *Endocrinology.* 2011;152(7):2626-33.

[192] Chen W, Leo S, Weng C, Yang X, Wu Y, Tang X. Mechanisms mediating renal sympathetic nerve activation in obesity-related hypertension. *Herz.* 2015;40(2):190-6.

[193] Kosari S, Rathner J, Badoer E. Central resistin enhances renal sympathetic nerve activity via phosphatidylinositol 3-kinase but reduces the activity to brown adipose tissue via extracellular signal-regulated kinase 1/2. *Journal of neuroendocrinology.* 2012;24 (11):1432-9.

[194] Barnes MA, Carson MJ, Nair MG. Non-traditional cytokines: how catecholamines and adipokines influence macrophages in immunity, metabolism and the central nervous system. *Cytokine.* 2015;72(2): 210-9.

[195] Guerrero-García JdJ, Carrera-Quintanar L, López-Roa RI, Márquez-Aguirre AL, Rojas-Mayorquín AE, Ortuño-Sahagún D. Multiple sclerosis and obesity: possible roles of adipokines. *Mediators of inflammation.* 2016;2016.

[196] Rodriguez-Pacheco F, Novelle M, Vazquez M, Garcia-Escobar E, Soriguer F, Rojo-Martinez G, García-Fuentes E, Malagon M, Dieguez

C. Resistin regulates pituitary lipid metabolism and inflammation *in vivo* and *in vitro*. *Mediators of inflammation.* 2013;2013.

[197] Asterholm IW, Rutkowski JM, Fujikawa T, Cho Y-R, Fukuda M, Tao C, Wang ZV, Gupta RK, Elmquist JK, Scherer PE. Elevated resistin levels induce central leptin resistance and increased atherosclerotic progression in mice. *Diabetologia.* 2014;57(6):1209-18.

[198] Singhal V, Misra M, Klibanski A. Endocrinology of anorexia nervosa in young people: recent insights. *Current opinion in endocrinology, diabetes, and obesity.* 2014;21(1):64.

[199] Challet E. Circadian aspects of adipokine regulation in rodents. *Best Practice & Research Clinical Endocrinology & Metabolism.* 2017.

In: Resistin
Editor: Gerald Maldonado
ISBN: 978-1-53614-543-4


*Chapter 2*

# RESISTIN AND EXERCISE

***Kazuhiko Kotani****
Division of Community and Family Medicine,
Jichi Medical University, Shimotsuke-City,
Tochigi, Japan

## ABSTRACT

Resistin was discovered as an adipocyte-driven molecule, and human resistin is predominantly secreted in the macrophages, with a positive connection with cardiometabolic pathophysiology. The biology of human resistin appears to differ from that of animal resistin, and the function of human resistin remains unknown. Using a MEDLINE/Pubmed database, we summarized the English literature investigating the effects of exercise intervention on the circulating resistin levels in humans. Regular exercise (regardless of modality) generally reduces the circulating resistin level, while aerobic/endurance exercise increases the level in healthy populations in particular. The summarized results and experimental/ clinical evidence indicate that resistin can have functions related to inflammation and oxidative stress. The circulating resistin level may be also useful as a marker of the effects of exercise. Further studies are needed.

* Corresponding Author Email: kazukotani@jichi.ac.jp.

**Keywords:** adipokine, aerobic exercise, resistance training, inflammation, oxidative stress

## RESISTIN AND ITS FUNCTION

Resistin, a 12.5-kDa cysteine-rich polypeptide, was discovered as an adipocyte-derived molecule with 'resistance' to insulin action in mice [1]. In rodents, this molecule is produced by adipocytes, linking to insulin resistance, while human resistin is mainly secreted by macrophages [2, 3]. Resistin is detectable in the circulation in humans but not in rodents [2, 3]. These differences imply the need for human-specific research to clarify the clinical relevance of resistin.

Although a positive association between resistin and cardiometabolic diseases has been suggested, the basic mechanisms of their association remain unclear [2, 3]. In macrophages, resistin induces the production of inflammatory cytokines [4, 5] and endothelial dysfunction [5, 6]. Inflammatory cytokines also induce the production of resistin [7]. In vascular cells, resistin induces oxidative stress responses [8]. Some clinical data have shown that the circulating resistin levels are higher under inflammatory/oxidative conditions than under adipocyte-dysregulated conditions [9, 10]. Further research to understand the function of resistin is required in humans.

## EFFECTS OF EXERCISE ON RESISTIN LEVELS

Exercise is the cornerstone for maintaining health and treating diseases, such as cardiometabolic pathologies [11, 12]. Exercise and physical activity are sometimes used interchangeably. Physical activity covers a broad range of bodily movement, while exercise refers to a planned physical activity, such as aerobic exercise (endurance exercise [e.g., running, jogging]) or resistance training (strength-developing exercise) [13]. In general, changes in the levels of certain molecules by exercise intervention are useful to find those molecules' functions [14], although the modality (i.e., type of exercise/training) as well as its intensity, frequency, and duration are all important determinants of the effects of exercise on such molecules.

We searched a MEDLINE/Pubmed database in a systematic review style through July 2018 and gathered the English literature (original articles) that investigated the effects of exercise intervention on the circulating resistin levels. The findings of the main relevant articles [15-28] were then summarized (Table 1).

With endurance exercise, such as a marathon, the resistin levels were increased, although only two studies explored this point and both had a single-arm design [15, 16].

With aerobic exercise, the resistin levels were reduced in populations with obesity-related diseases [17, 18, 20, 21, 23] and increased in healthy populations [19, 22]. This finding was consistent, even though only two studies (including one randomized controlled trial) were conducted in healthy populations [19, 22].

With resistance training, the resistin levels were basically reduced [24-26]. The training level in one study that showed no significant reduction in resistin was assumed to be very weak, leading to the lack of change in the resistin level [26].

With aerobic exercise plus resistance training, the resistin level was were reduced, although only one study explored this setting [18]. The change in the resistin level was slightly larger than that seen with aerobic exercise only [18]. With exercise (at the walking level) plus diet intervention, the resistin levels were unchanged [27, 28].

Taken together, these findings suggest that regular exercise can reduce the circulating resistin level, regardless of the modality. However, aerobic/endurance exercise can increase the level in healthy populations in particular.

### Table 1. Summarized results of resistin change following exercise intervention

| Authors | Design | Subjects (n, gender [M/W], age) | Exercise contents | Resistin change |
|---|---|---|---|---|
| *(Ultra-)marathon* | | | | |
| Roupas ND, et al. (ref. 15) | Single | Healthy marathon athletes (n = 17, M, 51.3 years) | Ultra-marathon (180 km of continuous running) | ↑ |
| Vuolteenaho K, et al. (ref. 16) | Single | Healthy marathoners (n = 46, M, 40.3 years) | Marathon (city marathon [42.2 km], 3.5 hours) | ↑ |
| *Aerobic exercise* | | | | |
| Kadoglou NP, et al. (ref. 17) | Control | Overweight/obese type 2 diabetes (M/W = 12/16, 56.8 years) | 16 weeks (4 days/week) | ↓ |
| Balducci S, et al. (ref. 18) | RCT | Type 2 diabetes and MetS (low-intensity: M/W = 11/9, 62.5 years; aerobic: M/W = 12/8, 64.3 years) | 12 weeks (low-intensity *vs.* aerobic [twice/week]) | ↓ |
| Rashidlamir A, et al. (ref. 19) | Control | Healthy (n = 15, M, 40.1 years) | 8 weeks (4 days/week) | ↑ |
| Wenning P, et al. (ref. 20) | Single | Overweight type 2 diabetes (n = 14, M, 61.0 years) | 12 weeks (twice/week) | ↓ |
| Many G, et al. (ref. 21) | Single | Morbidly obese teens (M/W = 2/9,15.0 years) | 8 weeks (moderate-intensity [gradual increase/week]) | ↓ |
| Ozcan O, et al. (ref. 22) | RCT | Healthy (aerobic: n = 20, W, 33.2 years; core: n = 20, W, 34.9 years) | 16 weeks (aerobic or core training [4 days/week]) | ↑ (both groups) |
| Lee SS, et al. (ref. 23) | Control | Obese (n = 10, M/W = ?, 16.7 years), type 2 diabetes (n = 10, M/W = ?,16.8 years) | 12 weeks (4 days/week) | ↓ (both groups) |

| **Authors** | **Design** | **Subjects (n, gender [M/W], age)** | **Exercise contents** | **Resistin change** |
|---|---|---|---|---|
| *Resistance training* | | | | |
| Prestres J, et al. (ref. 24) | Single | Sedentary (n = 35, W, 63.2 years) | 16 weeks (periodized) | ↓ |
| Botero JP, et al. (ref. 25) | Single | Postmenopausal (n = 23, W, 63.0 years) | 12 months (twice/week) | ↓ |
| Lubkowska A, et al. (ref. 26) | Single | Obese (n = 30, M, 40.0 years) | 6 months (whole-body cryostimulation [5 days/week]) | → |
| *Aerobic+resistance training* | | | | |
| Balducci S, et al. (ref. 18) | RCT | Type 2 diabetes and MetS (low-intensity: M/W = 11/9, 62.5 years; aerobic +resistance training: M/W = 14/8, 60.6 years) | 12 weeks (low-intensity *vs.* aerobic + resistance [twice/week]) | ↓ |
| *Exercise+diet* | | | | |
| Oh EG, et al. (ref. 27) | RCT | MetS (n = 52, W, 62.7 years [control; n = 21]) | 6 months (yoga, aerobic+low-energy diet) | → |
| Giannopoulou I, et al. (ref. 28) | RCT | Postmenopausal type 2 diabetes (n = 33, W, 50 - 70 years) | 14 weeks (diet or exercise *vs.* exercise + diet) | → |

M/W: men/women (n), RCT: randomized controlled trial, MetS: metabolic syndrome.

## The Possible Function of Resistin

Based on the summarized results regarding the changes in the resistin level with exercise intervention and experimental/clinical evidence [2-10], we propose that resistin has a possible function associated with inflammation and oxidative stress. The resistin molecule can be mainly involved in inflammation and oxidative stress as a causative factor (partly a consequent factor following inflammation and oxidative stress) [4, 5, 7, 8]. The circulating resistin levels are increased under inflammatory/ oxidative conditions, including cardiometabolic disease, in humans (the levels are positively correlated with C-reactive protein or interleukin-6 and, unexpectedly, are not always correlated with adipose-dysregulated disorders) [9, 10]; as such, resistin can be a positive biomarker for inflammation and oxidative stress. In general, exercise reduces inflammation and oxidative stress, with the reduced production of inflammatory/oxidative molecules and increased production of anti-inflammatory mediators and anti-oxidants (this is the defense system) [18, 19, 29, 30]. This may explain in part the reduction in the resistin level with exercise. Also, as the baseline level of resistin is high in populations with cardiometabolic diseases [9, 10], resistin may be markedly reduced by exercise in those populations.

Possible explanations for the increase in the resistin level induced by endurance/aerobic exercise and in healthy populations are as follows: hypercortisolism induced by endurance exercise level may increase the amount of resistin mRNA [7]. Also, by case, endurance exercise induces harmful inflammation/oxidative stress in healthy populations [30], which may contribute to the increased levels of resistin.

## Resistin as an Exercise-Related Marker

Even though the circulating resistin level is often influenced by the exercise situation and study population, a reduction in the resistin level presumably indicates the alleviation of inflammatory/oxidative conditions by exercise. No change in the resistin level might be suggestive of weak or ineffective intervention of exercise [26]. Thus, we think that resistin can be a marker reflective of the effects of exercise.

Recently, a marathon study found that resistin was a novel marker of cartilage degradation, due to the positive relationship between the resistin level and cartilage-related measurements [16]. This is an interesting topic concerning the function of resistin. Further research will help clarify the marker role of resistin in relation to exercise.

## REFERENCES

[1] Steppan, C. M., Bailey, S. T., Bhat, S., Brown, E. J., Banerjee, R. R., Wright, C. M., Patel, H. R., Ahima, R. S., Lazar, M. A. The hormone resistin links obesity to diabetes. *Nature,* 2001; 409:307 - 12.

[2] Barnes, K. M., Miner, J. L. Role of resistin in insulin sensitivity in rodents and humans. *Curr. Protein Pept. Sci.*, 2009;10:96 - 107.

[3] Park, H. K., Ahima, R. S. Resistin in rodents and humans. *Diabetes Metab. J.,* 2013;37:404 - 14.

[4] Lehrke, M., Reilly, M. P., Millington, S. C., Iqbal, N., Rader, D. J., Lazar, M. A. An inflammatory cascade leading to hyperresistinemia in humans. *PLoS Med.,* 2004;1:e45.

[5] Burnett, M. S., Lee, C. W., Kinnaird, T. D., Stabile, E., Durrani, S., Dullum, M. K., Devaney, J. M., Fishman, C., Stamou, S., Canos, D., Zbinden, S., Clavijo, L. C., Jang, G. J., Andrews, J. A., Zhu, J., Epstein, S. E. The potential role of resistin in atherogenesis. *Atherosclerosis*, 2005;182:241 - 8.

[6] Verma, S., Li, S. H., Wang, C. H., Fedak, P. W., Li, R. K., Weisel, R. D., Mickle, D. A. Resistin promotes endothelial cell activation: further evidence of adipokine-endothelial interaction. *Circulation,* 2003; 108:736 - 40.

[7] Stejskal, D., Adamovská, S., Bartek, J., Juráková, R., Prosková, J. Resistin-concentrations in persons with type 2 diabetes mellitus and in individuals with acute inflammatory disease. *Biomed. Pap. Med. Fac. Univ. Palacky Olomouc Czech Repub.*, 2003;147:63 - 9.

[8] Chen, C., Jiang, J., Lü, J. M., Chai, H., Wang, X., Lin, P. H., Yao, Q. Resistin decreases expression of endothelial nitric oxide synthase through oxidative stress in human coronary artery endothelial cells. *Am. J. Physiol. Heart Circ. Physiol.*, 2010;299:H193 - 201.

[9] Qi, Q., Wang, J., Li, H., Yu, Z., Ye, X., Hu, F. B., Franco, O. H., Pan, A., Liu, Y., Lin, X. Associations of resistin with inflammatory and fibrinolytic markers, insulin resistance, and metabolic syndrome in middle-aged and older Chinese. *Eur. J. Endocrinol.*, 2008; 159: 585 - 93.

[10] Laurikka, A., Vuolteenaho, K., Toikkanen, V., Rinne, T., Leppänen, T., Tarkka, M., Laurikka, J., Moilanen, E. Adipocytokine resistin correlates with oxidative stress and myocardial injury in patients undergoing cardiac surgery. *Eur. J. Cardiothorac. Surg.*, 2014; 46: 729 - 36.

[11] Puffer, J. C. Exercise and heart disease. *Clin. Cornerstone,* 2001; 3:1 - 9.

[12] Montesi, L., Moscatiello, S., Malavolti, M., Marzocchi, R., Marchesini, G. Physical activity for the prevention and treatment of metabolic disorders. *Intern. Emerg. Med.*, 2013; 8:655 - 66.

[13] Johannsen, N. M., Swift, D. L., Lavie, C. J., Earnest, C. P., Blair, S. N., Church, T. S. Combined aerobic and resistance training effects on glucose homeostasis, fitness, and other major health indices: A review of current guidelines. *Sports Med.*, 2016; 46:1809 - 18.

[14] Palmefors, H., DuttaRoy, S., Rundqvist, B., Börjesson, M. The effect of physical activity or exercise on key biomarkers in atherosclerosis-a systematic review. *Atherosclerosis,* 2014; 235:150 - 61.

[15] Roupas, N. D., Mamali, I., Maragkos, S., Leonidou, L., Armeni, A. K., Markantes, G. K., Tsekouras, A., Sakellaropoulos, G. C., Markou, K. B., Georgopoulos, N. A. The effect of prolonged aerobic exercise on

serum adipokine levels during an ultra-marathon endurance race. *Hormones (Athens),* 2013; 12:275 - 82.

[16] Vuolteenaho, K., Leppänen, T., Kekkonen, R., Korpela, R., Moilanen, E. Running a marathon induces changes in adipokine levels and in markers of cartilage degradation--novel role for resistin. *PLoS One*, 2014;9:e110481.

[17] Kadoglou, N. P., Perrea, D., Iliadis, F., Angelopoulou, N., Liapis, C., Alevizos, M. Exercise reduces resistin and inflammatory cytokines in patients with type 2 diabetes. *Diabetes Care,* 2007; 30:719 - 21.

[18] Balducci, S., Zanuso, S., Nicolucci, A., Fernando, F., Cavallo, S., Cardelli, P., Fallucca, S., Alessi, E., Letizia, C., Jimenez, A., Fallucca, F., Pugliese, G. Anti-inflammatory effect of exercise training in subjects with type 2 diabetes and the metabolic syndrome is dependent on exercise modalities and independent of weight loss. *Nutr. Metab. Cardiovasc. Dis.,* 2010; 20:608 - 17.

[19] Rashidlamir, A., *Saadatnia,* A. The effect of eight weeks of aerobic training on the plasma level of adiponectin, leptin, and *resistin* **i**n healthy middle-aged men. *Sci. Sports, 2012*; 27 351 - 6.

[20] Wenning, P., Kreutz, T., Schmidt, A., Opitz, D., Graf, C., Voss, S., Bloch, W., Brixius, K. Endurance exercise alters cellular immune status and resistin concentrations in men suffering from non-insulin-dependent type 2 diabetes. *Exp. Clin. Endocrinol. Diabetes,* 2013; 121:475 - 82.

[21] Many, G., Hurtado, M. E., Tanner, C., Houmard, J., Gordish-Dressman, H., Park, J. J., Uwaifo, G., Kraus, W., Hagberg, J., Hoffman, E. Moderate-intensity aerobic training program improves insulin sensitivity and inflammatory markers in a pilot study of morbidly obese minority teens. *Pediatr. Exerc. Sci.,* 2013; 25:12 - 26.

[22] Ozcan, O., Bostanci, M. Ö., Cicek, G., Yamaner, F. The effects of two different exercise programmes on adipose tissue hormones in sedentary middle-aged women. *Arch. Physiol. Biochem.,* 2015; 121: 50 - 5.

[23] Lee, S. S., Kang, S. Effects of regular exercise on obesity and type 2 diabetes mellitus in Korean children: improvements glycemic control and serum adipokines level. *J. Phys. Ther. Sci.,* 2015; 27:1903 - 7.

[24] Prestes, J., Shiguemoto, G., Botero, J. P., Frollini, A., Dias, R., Leite, R., Pereira, G., Magosso, R., Baldissera, V., Cavaglieri, C., Perez, S. Effects of resistance training on resistin, leptin, cytokines, and muscle force in elderly post-menopausal women. *J. Sports Sci.,* 2009; 27: 1607 - 15.

[25] Botero, J. P., Shiguemoto, G. E., Prestes, J., Marin, C. T., Do Prado, W. L., Pontes, C. S., Guerra, R. L., Ferreia, F. C., Baldissera, V., Perez, S. E. Effects of long-term periodized resistance training on body composition, leptin, resistin and muscle strength in elderly post-menopausal women. *J. Sports Med. Phys. Fitness*, 2013; 53:289 - 94.

[26] Lubkowska, A., Dudzińska, W., Bryczkowska, I., Dołęgowska, B. Body composition, lipid profile, acipokine concentration, and antioxidant capacity changes during interventions to treat overweight with exercise programme and whole-body cryostimulation. *Oxid. Med. Cell Longev.*, 2015; 2015:803197.

[27] Oh, E. G., Bang, S. Y., Kim, S. H., Hyun, S. S., Chu, S. H., Jeon, J. Y., Im, J. A., Lee, J. E., Lee, M. K. Therapeutic lifestyle modification program reduces plasma levels of the chemokines CRP and MCP-1 in subjects with metabolic syndrome. *Biol. Res. Nurs.,* 2013; 15: 48 - 55.

[28] Giannopoulou, I., Fernhall, B., Carhart, R., Weinstock, R. S, Baynard, T., Figueroa, A., Kanaley, J. A. Effects of diet and/or exercise on the adipocytokine and inflammatory cytokine levels of postmenopausal women with type 2 diabetes. *Metabolism,* 2005; 54:866 - 75.

[29] Kotani, K., Koibuchi, H., Yamada, T., Taniguchi, N. The effects of lifestyle modification on a new oxidized low-density lipoprotein marker, serum amyloid A-LDL, in subjects with primary lipid disorder. *Clin. Chim. Acta*, 2009; 409:67 - 9.

[30] Michigan, A., Johnson, T. V., Master, V. A. Review of the relationship between C-reactive protein and exercise. *Mol. Diagn. Ther.*, 2011; 15:265 - 75.

[31] Simioni, C., Zauli, G., Martelli, A. M., Vitale, M., Sacchetti, G., Gonelli, A., Neri, L. M. Oxidative stress: role of physical exercise and antioxidant nutraceuticals in adulthood and aging. *Oncotarget,* 2018; 9:17181 - 98.

In: Resistin
Editor: Gerald Maldonado
ISBN: 978-1-53614-543-4


*Chapter 3*

# RESISTIN AND DIABETES

***Zoran Gluvic[1], Milan Obradovic[2], Predrag Popovic[3], Bratislav Stankovic[4] and Esma R. Isenovic[2]***
[1]Zemun Clinical Hospital, School of Medicine, University of Belgrade, Department of Endocrinology, Diabetes and ICU, Zemun, Serbia
[2]Institute for Nuclear Sciences "Vinca," University of Belgrade, Laboratory of Radiobiology and Molecular Genetics, Belgrade, Serbia
[3]Ultrasound diagnostic clinic "HRAST DR POPOVIĆ", Belgrade, Serbia
[4]Veterinary Clinic "dr Maja Ilic," Belgrade, Serbia

## ABSTRACT

Resistin is a peptide that belongs to resistin-like molecules family. In humans, it is mainly produced from macrophages, bone marrow cells, and peripheral blood mononuclear cells. Resistin promotes inflammation via up-regulation of inflammatory cytokines' production and secretion especially in some proinflammatory conditions, such as obesity, metabolic syndrome and type 2 diabetes mellitus (T2DM). Furthermore, chronic hyperresistinemiais strongly is associated with inflammation, which is the key factor in T2DM pathogenesis. By the inflammation favoring, the increased resistin levels worsen endothelial dysfunction and promote earlier onset of diabetes vascular complications. Also, resistin can promote vasculopathy through the promotion of macrophage scavenger receptors and CD36 atheroma macrophages expression, the formation of foam cells, as well as proliferation and migration of vascular smooth muscle cells. This favours endothelial dysfunction, which results in limited nitric oxide and increased superoxide production. Some studies revealed the correlation of higher resistin levels with proatherogenic lipids, such as very low-density lipoproteins, Apo B, triglycerides, and small dense low-density lipoproteins, especially in T2DM patients. In this chapter, we have presented recent literature data related to the link between the resistin and diabetes.

**Keywords:** resistin, inflammation, diabetes

## INTRODUCTION

Diabetes mellitus (DM) is one of the most common chronic diseases estimated to affect 425 million individuals worldwide [1]. DM is a group of metabolic disorders, characterized by hyperglycaemia and disturbing level or action of insulin. Individuals with diabetes suffer from high morbidity and mortality rates because of the development of microvascular, macro-vascular, and neuropathic complications [2].

A variety of adipokines that are involved in the regulation of glucose homeostasis also have been linked with the development of inflammation, insulin resistance, and DM. Since it was discovered, resistin is suspected to be involved in the development and progression of insulin resistance and DM [3-5]. Resistin promotes inflammation via up-regulation of inflammatory cytokines' production and secretion especially in some proinflammatory conditions, such as obesity, metabolic syndrome and types 2 DM (T2DM) [6-8]. The hyperresistinemia worsen endothelial dysfunction and promote earlier onset of diabetes vascular complications through inflammation favouring [9, 10]. Also, resistin can promote vasculopathy through the promotion of macrophage scavenger receptors and CD36 atheroma macrophages expression, the formation of foam cells, as well as proliferation and migration of vascular smooth muscle cells

[11, 12]. This favours endothelial dysfunction, which results in limited nitric oxide and increased superoxide production [13, 14]. In this chapter, we review the literature data related to the role of resistin, as an inflammatory cytokine, in the development of DM.

## RESISTIN

Resistin is a peptide (Mw 12.5 kDa) that belongs to the family of resistin-like molecules (RELM), a class of cysteine-rich proteins that forms multimeric complexes [5, 15, 16]. Resistin was discovered in 2000-2001 simultaneously by three different research groups [3, 17, 18]. The name resistin come from its ability to resist (interfere with) insulin action [3]. The other synonyms for resistin are an adipose secretory factor (ADSF) [17], and cysteine-rich secreted protein found in inflammatory zone (FIZZ3) [18], which are related to the way of its discovery, and are rarely in use.

In humans and animal models (mice), there are different mechanisms of regulation, tissue distribution and functions of resistin [3, 5, 19-21]. The human resistinis mainly produced in peripheral blood mononuclear cells (PBMC), macrophages and bone marrow cells [5, 22], while in mice resistin is predominantly expressed in white adipose tissue [3]. Also, the human gene of resistin is localized on chromosome 19, while in the mouse on chromosome 8 [19, 23]. In circulation, resistin tends to form multimeric complexes, such as a high-molecular weight (HMW) form, and a low-molecular weight (LMW) form [5, 24, 25]. In contrast to mice, resistin in human tends to form oligomers that are biologically more active [26, 27].

Although studies on mouse model showed that resistin is involved in the pathogenesis of insulin resistance and DM, considering all differences between human and murine resistin arise the question is it reliable to extrapolate data from rodent studies on human [5, 28]. Also, human studies (not all) determined the increased level of resistin in serum and resistin mRNA level in PBMC of patients with DMT2 [8, 29], but the role of resistin in the pathogenesis of DM in humans still to be elucidated. Resistin exerts pleiotropic action and has been involved in different pathophysiological processes (not only diabetes) such as inflammation, atherosclerosis, cardiovascular diseases, non-alcoholic fatty liver disease, malignancy, and chronic kidney diseases [4, 5].

## DIABETES

Diabetes mellitus is a chronic, progressive, polygenic disease, which occurs as a result of insufficiency or inadequate activity of insulin [30, 31]. DM represents a risk factor for the development of cardiovascular diseases, and it is one of the five leading causes of death in most developed countries [2]. The main characteristics of DM are hyperglycaemia and different metabolic disorders that lead to long-term complications such as heart disease, stroke, nephropathy, retinopathy, peripheral neuropathy and lower limb amputation [2, 31]. Also, untreated diabetes leads to diabetic ketoacidosis, polyuria, polydipsia, weight loss and in extreme cases diabetic coma (mortality rate of 7%) [2, 31].

There is a different classification of DM based on the type of treatment or pathogenesis mechanisms, but primarily it can be classified to type 1 diabetes (DMT1), DMT2 and gestational diabetes [30, 31]. DMT1 or insulin-dependent diabetes represents only 8% of the total number of patients with DM and usually peaks in adolescence, but it can occur at any age [32]. Generally, it is considered that this form of DM occurs in genetically predisposed individuals as a result of toxins or infectious injuries of pancreatic β cells associated with the effects of an aggressive immune system that continuously destroys β-cells [32]. In contrast, DMT2 begins later in life, and it is the most common form of DM, present in 85-90% of patients with DM. Aetiology of DMT2 is multifactorial, but the significant influence of its manifestation has risk factors including inadequate diet, obesity and physical inactivity [31]. DMT2 begins with insulin resistance, a state characterized by the decreased response of cells to the insulin action [33, 34]. However, in the later stage of disease, patients with DMT2 become insulin-dependent for blood glucose control [35]. DMT2 is associated with significant complications, particularly in the vascular system [2].

Gestational diabetes is developing in women during pregnancy [30, 31]. Although an increased risk of complications during pregnancy, gestational diabetes also increased the risk of DMT2 in the future, for both mother and children [30, 31].

## RESISTIN AND DIABETES

Insulin resistance is one of the major factors that play role in the pathogenesis of DMT2. As a critical signal molecule [36], resistin functions as the link between obesity, DMT2 and insulin resistance [37]. More precisely, resistin coordinates between energy storage organ (adipose tissue) and insulin-target tissues (such as liver, muscles and fat) [38]. Lee et al. presented adenylyl cyclase-associated protein 1 (CAP1) as a receptor for human resistin. Binding of human resistin to monocytes' adenylyl cyclase-associated protein 1 increases cyclic adenosine monophosphate and protein kinase A activity, and nuclear factor kappa B-related transcription of inflammatory cytokines, as well [39]. Resistin has been shown to be expressed by human pancreatic islets cells, and its expression has been up-regulated in DMT2 [40, 41]. Resistin co-localizes pancreatic islet cells with insulin of both healthy and diabetic subjects. However, the degree of co-localization was higher in diabetic patients [42]. Activation of resistin in the pancreatic islet cells inhibits cell surface glucose transporters and insulin signaling [41, 43], with subsequent β-cell loss enhanced by proinflammatory mechanism [41, 44, 45].

Resistin correlated with insulin resistance in patients with the metabolic syndrome (MetS) or DMT2 [46, 47]. These correlations were consistent in women, but not in men. Possible gender-specific differences in adipose tissue depots responsible for resistin production in females or males may explain these findings [48]. Numerous studies were performed to unravel the effects of resistin on the DMT2 pathogenesis. Increase in circulating inflammatory adipokines has been suggested as predictive of DMT2 [49]. Some studies showed that circulating levels of resistin and visfatin are upregulated in DMT2 [45, 50-3]. As inflammation plays a crucial role in the pathogenesis of DMT2, resistin could be implicated in the development of DMT2 via induction and promotion of inflammation or insulin resistance [54-6].

Rodent studies indicate that resistin regulates insulin receptor substrate 1 function by affecting its tyrosine phosphorylation, reduces insulin-stimulated glucose uptake in skeletal muscles and decreases GLUT-4 (glucose transporter) translocation following insulin stimulation [57, 58]. Another promising hypothesis regarding resistin influence on insulin resistance and DMT2 was resistin-induced gluconeogenesis increase in the liver in patients suffered from DMT2 [59]. Rodent study failed to prove it [60]. Surprisingly, metformin treatment in animals corrects hyperglycemia and insulin resistance but increases resistin gene expression at the protein level [60].

Disruption of the insulin signaling in insulin target tissues is the basis of insulin resistance induced by resistin [61, 62]. Resistin plays a crucialrole in angiogenesis-related vascular disorders. It is evidenced by resistin effects on proliferation, migration and activation of endothelial cells, as well an increase in expression of vascular endothelial growth factor receptors, matrix metalloproteinases and activation of extracellular signal-regulated protein kinases 1 and 2and p38 pathways [63]. In DMT2 patients, resistin positively correlated with inflammatory markers, such as C-reactive protein (CRP) or interleukin 6 (IL-6) [48, 64, 65]. Macrophages that infiltrate the stroma of the adipose tissue and/or the vascular endothelium can locally produce and secrete resistin in addition to IL-13, tumour necrosis factor-α and IL-6, which may enhance the inflammatory load inducing insulin resistance and vascular dysfunction in DMT2 patients [66, 67]. Hyper resistenemia is associated with micro- and macrovascular DMT2 complications [65, 68-70]. It has been reported that elevated resistin levels contribute to hypertension [71] and to accelerate DMT2-associated microangiopathy [72]. It seems to be that resistin-induced micro-angiopathies are induced by the co-existence of endothelial dysfunction and oxidative stress in DMT2 [73, 74]. By joining to dyslipidemia and hypertension, hyperresistenemia contributes to DMT2-accelerated large vessel diseases (i.e., cerebro-, cardio-, and peripheral vascular disease) by chronic inflammation [75-77]. In patients with DMT2, serum resistin level can serve as a marker of both coronary artery disease and restenosis [65]. Additionally, hyperresistenemia is a predictive marker for major adverse cardiac events and plaque instability in already established coronary artery disease not just in DMT2 patients [78-79]. However, there are studies that failed to show the link between elevated resistin levels and large vessel diseases [80, 81].

Novel genetic and polymorphism population-based studies showed that single nucleotide polymorphisms (SNPs) of the resistin gene (RETN) are associated with the development of DMT2, such as SNPs at position +299(G/A) [82, 83]. Another study showed that SNP 420C/G in the promoter region of RETN plays an important role in resistin gene expression and determination of plasma resistin concentration in humans [83]. Some SNPs of RETN may serve as susceptibility predictive marker to DMT2 in offspring of diabetics. Additionally, hyperesistenemia could predict mortality risk in DMT2 patients and can be associated with deterioration of the natural history of DMT2 [83, 84]. Some studies showed no significant difference in resistin levels in obese and slim diabetics compare to slim, healthy control [85, 86], between obese diabetics and non-diabetics [87, 88], and between normal weight and overweight persons [89]. Additionally, some studies showed the difference in resistin levels between patients with and with no diabetes [90], between obese diabetics and obese non-diabetics and between obese diabetics, obese non-diabetics and normal weight non-diabetic controls [90, 91].

## CONCLUSION

Despite extensive research in the field, the relationship between resistin, hyperglycemia or insulin resistance is still conflicting. At the pancreatic level, inhibition of resistin local production and effects could probably affect inflammation-induced beta cell loss. On the systemic level, resistin inhibition could decrease inflammatory mediators' syntheses, low-grade inflammation, endothelial dysfunction, and subsequent acceleration and extension of the atherosclerosis process. Cumbersome clinical importance of those abovecan be seen through the frequent presence of micro- and macrovascular complications, which can dramatically change the course of DMT2. New efforts must be introduced to elucidate the significance of resistin level decrease regarding the prediction of insulin resistance, DMT2 or even DMT1. Further population-based, translational and interventional studies are needed to unravel clinical significance of some SNPs in RETN gene, the influence of resistin on liver gluconeogenesis, and the effects of some new drugs on such process. That step can be the platform for new management strategies for diabetes patients.

## Acknowledgments

This work has been supported by grant 173033 (to E.R.I.) from the Ministry of Education, Science and Technological Development, Republic of Serbia.

## References

[1] *IDF Diabetes Atlas*. 8th edition. Available at: http://www.diabetes atlas.org/.2017.

[2] Deshpande AD, Harris-Hayes M, Schootman M. Epidemiology of diabetes and diabetes-related complications. *Phys Ther.* 2008;88:1254-64.

[3] Steppan CM, Bailey ST, Bhat S, Brown EJ, Banerjee RR, Wright CM, Patel HR, Ahima RS, Lazar MA. The hormone resistin links obesity to diabetes. *Nature*. 2001;409:307-12.

[4] Filkova M, Haluzik M, Gay S, Senolt L. The role of resistin as a regulator of inflammation: Implications for various human pathologies. *Clin Immunol*. 2009;133:157-70.

[5] Jamaluddin MS, Weakley SM, Yao Q, Chen C. Resistin: functional roles and therapeutic considerations for cardiovascular disease. *Br J Pharmacol.* 2012;165:622-32.

[6] Weisberg SP, McCann D, Desai M, Rosenbaum M, Leibel RL, Ferrante AW, Jr. Obesity is associated with macrophage accumulation in adipose tissue. *J Clin Invest.* 2003;112:1796-808.

[7] Baldasseroni S, Urso R, Maggioni AP, Orso F, Fabbri G, Marchionni N, Tavazzi L. Prognostic significance of serum uric acid in outpatients with chronic heart failure is complex and related to body mass index: data from the IN-CHF Registry. *Nutr Metab CardiovascDis.* 2012;22:442-8.

[8] Gharibeh MY, Al Tawallbeh GM, Abboud MM, Radaideh A, Alhader AA, Khabour OF. Correlation of plasma resistin with obesity and insulin resistance in type 2 diabetic patients. *Diabetes & Metabolism.* 2010;36:443-9.

[9] Verma S, Li SH, Wang CH, Fedak PW, Li RK, Weisel RD, Mickle DA. Resistin promotes endothelial cell activation: further evidence of adipokine-endothelial interaction. *Circulation.* 2003;108:736-40.

[10] Kawanami D, Maemura K, Takeda N, Harada T, Nojiri T, Imai Y, Manabe I, Utsunomiya K, Nagai R. Direct reciprocal effects of resistin and adiponectin on vascular endothelial cells: a new insight into adipocytokine-endothelial cell interactions. *Biochem Biophys Res Commun.* 2004;314:415-9.

[11] Jung HS, Park KH, Cho YM, Chung SS, Cho HJ, Cho SY, et al. Resistin is secreted from macrophages in atheromas and promotes atherosclerosis. *Cardiovasc Res.* 2006;69:76-85.

[12] Lee SE, Kim HS. Human resistin in cardiovascular disease. *J SmoothMuscle Res.* 2012;48:27-35.

[13] Maenhaut N, Van de Voorde J. Regulation of vascular tone by adipocytes. *BMC Med.* 2011;9:25.

[14] Chen C, Jiang J, Lu JM, Chai H, Wang X, Lin PH, Yao Q. Resistin decreases expression of endothelial nitric oxide synthase through oxidative stress in human coronary artery endothelial cells. *Am J Physiol Heart Circ Physiol.* 2010;299: H193-201.

[15] Kontrogianni-Konstantopoulos A. Resisting resistin; it good for the heart. *Journal of molecular and cellular cardiology*. 2011;51:141-3.
[16] Steppan CM, Brown EJ, Wright CM, Bhat S, Banerjee RR, Dai CY, et al. A family of tissue-specific resistin-like molecules. *Proc Natl Acad Sci US*. 2001;98:502-6.
[17] Kim KH, Lee K, Moon YS, Sul HS. A cysteine-rich adipose tissue-specific secretory factor inhibits adipocyte differentiation. *J Biol Chem.* 2001;276:11252-6.
[18] Holcomb IN, Kabakoff RC, Chan B, Baker TW, Gurney A, Henzel W, et al. FIZZ1, a novel cysteine-rich secreted protein associated with pulmonary inflammation, defines a new gene family. *TheEMBO journal*. 2000;19:4046-55.
[19] Ghosh S, Singh AK, Aruna B, Mukhopadhyay S, Ehtesham NZ. The genomic organization of mouse resistin reveals major differences from the human resistin: functional implications. *Gene.* 2003;305:27-34.
[20] Yang RZ, Huang Q, Xu A, McLenithan JC, Eisen JA, Shuldiner AR, Alkan S, Gong DW. Comparative studies of resistin expression and phylogenomics in human and mouse. *Biochem Biophys Res Commun.* 2003;310:927-35.
[21] Koerner A, Kratzsch J, Kiess W. Adipocytokines: leptin--the classical, resistin--the controversial, adiponectin--the promising, and more to come. *Best Pract Res Clin Endocrinol Metab*. 2005;19:525-46.
[22] Patel L, Buckels AC, Kinghorn IJ, Murdock PR, Holbrook JD, Plumpton C, Macphee CH, Smith SA. Resistin is expressed in human macrophages and directly regulated by PPAR gamma activators. *Biochem Biophys Res Commun*. 2003;300:472-6.
[23] Steppan CM, Lazar MA. The current biology of resistin. *J Intern Med.* 2004;255:439-47.
[24] Patel SD, Rajala MW, Rossetti L, Scherer PE, Shapiro L. Disulfide-dependent multimeric assembly of resistin family hormones. *Science.* 2004;304:1154-8.
[25] Chen J, Wang L, Boeg YS, Xia B, Wang J. Differential dimerization and association among resistin family proteins with implications for

functional specificity. *The Journal of endocrinology*. 2002;175:499-504.

[26] Aruna B, Islam A, Ghosh S, Singh AK, Vijayalakshmi M, Ahmad F, Ehtesham NZ. Biophysical analyses of human resistin: oligomer formation suggests novel biological function. *Biochemistry.* 2008;47:12457-66.

[27] Aruna B, Ghosh S, Singh AK, Mande SC, Srinivas V, Chauhan R, Ehtesham NZ. Human recombinant resistin protein displays a tendency to aggregate by forming intermolecular disulfide linkages. *Biochemistry.* 2003;42:10554-9.

[28] Abate N, Sallam HS, Rizzo M, Nikolic D, Obradovic M, Bjelogrlic P, Isenovic ER. Resistin: an inflammatory cytokine. Role in cardiovascular diseases, diabetes and the metabolic syndrome. *Curr Pharm Des.* 2014;20:4961-9.

[29] Tsiotra PC, Tsigos C, Anastasiou E, Yfanti E, Boutati E, Souvatzoglou E, Kyrou I, Raptis SA. Peripheral mononuclear cell resistin mRNA expression is increased in type 2 diabetic women. *Mediators Inflamm.* 2008;892864:21.

[30] Diagnosis and classification of diabetes mellitus. *Diabetes Care.* 2010;33:dc10-S062.

[31] Association AD. Classification and Diagnosis of Diabetes. *Diabetes Care.* 2015;38:S8-S16.

[32] van Belle TL, Coppieters KT, von Herrath MG. Type 1 diabetes: etiology, immunology, and therapeutic strategies. *Physiol Rev.* 2011;91:79-118.

[33] Kahn SE, Hull RL, Utzschneider KM. Mechanisms linking obesity to insulin resistance and type 2 diabetes. *Nature.* 2006;444:840-6.

[34] Fujimoto WY. The importance of insulin resistance in the pathogenesis of type 2 diabetes mellitus. *Am J Med.* 2000;17:9S-14S.

[35] Nambam B, Aggarwal S, Jain A. Latent autoimmune diabetes in adults: A distinct but heterogeneous clinical entity. *World J Diabetes.* 2010;1:111-5.

[36] SaeediBorujeni MJ, Esfandiary E, Taheripak G, Codoñer-Franch P, Alonso-Iglesias E, Mirzaei H. Molecular aspects of diabetes mellitus:

Resistin, microRNA, and exosome. *J Cell Biochem* 2018;119:1257-72.

[37] Santilli F, Liani R, Di Fulvio P, Formoso G, Simeone P, Tripaldi R et al. Increased circulating resistin is associated with insulin resistance, oxidative stress and platelet activation in type 2 diabetes mellitus. *Thromb Haemost* 2016;116:1089-99.

[38] Codoñer-Franch P, Alonso-Iglesias E. Resistin: insulin resistance to malignancy. *Clin Chim Acta* 2015;438:46-54.

[39] Lee S, Lee HC, Kwon YW, Lee SE, Cho Y, Kim J et al. Adenylyl cyclase-associated protein 1 is a receptor for human resistin and mediates inflammatory actions of human monocytes. *Cell Metab* 2014;19:484-97.

[40] Minn AH, Patterson NB, Pack S, Hoffmann SC, Gavrilova O, Vinson C et al. Resistinis expressed in pancreatic islets. *Biochim BiophysRes Commun* 2003;310:641-5.

[41] Dunmore SJ1, Brown JE. The role of adipokines in β-cell failure of type 2 diabetes. *J Endocrinol* 2013;216:T37-45.

[42] Al-Salam S, Rashed H, Adeghate E. Diabetes mellitus is associated with an increased expression of resistin in human pancreatic islet cells. *Islets* 2011;3:246-9.

[43] Kim SJ, Nian C, McIntosh CH. Resistin knockout mice exhibit impaired adipocyte glucose-dependent insulinotropic polypeptide receptor (GIPR) expression. *Diabetes* 2013;62:471-7.

[44] Kusminski CM, McTernan PG, Kumar S. Role of resistin in obesity, insulin resistance and Type II diabetes. *Clin Sci* (Lond) 2005;109:243-56.

[45] Stofkova A. Resistin and visfatin: regulators of insulin sensitivity, inflammation and immunity. *Endocr Regul* 2010;44:25-36.

[46] Asano H, Izawa H, Nagata K, Nakatochi M, Kobayashi M, Hirashiki A et al. Plasma resistin concentration determined by common variants in the resistin gene and associated with metabolic traits in an aged Japanese population. *Diabetologia* 2010;53:234-46.

[47] Malo E, Ukkola O, Jokela M, Moilanen L, Kähönen M, Nieminen MS et al. Resistin is an indicator of the metabolic syndrome according to

five different definitions in the Finnish Health 2000 survey. *Metab Syndr Relat Disord* 2011;9:203-10.

[48] de Luis DA, Gonzalez Sagrado M, Conde R, Aller R, Izaola O, Perez Castrillon JL et al. Relation of resistin levels with cardiovascular risk factors and insulin resistance in non-diabetes obese patients. *Diabetes Res Clin Pract* 2009;84:174-8.

[49] Srinivasan M, Meadows ML, Maxwell L. Assessment of Salivary Adipokines Resistin, Visfatin, and Ghrelin as Type 2 Diabetes Mellitus Biomarkers. *Biochemistry Research International* 2018; 2018:7463796.

[50] Li L, Yang G, Li Q, Tang Y, Yang M, Yang H et al. Changes and relations of circulating visfatin, apelin, and resistin levels in normal, impaired glucose tolerance, and type 2 diabetic subjects. *Exp Clin Endocrinol Diabetes* 2006;114:544-8.

[51] Hasegawa G, Ohta M, Ichida Y, Obayashi H, Shigeta M, Yamasaki M et al. Increased serum resistin levels in patients with type 2 diabetes are not linked with markers of insulin resistance and adiposity. *Acta Diabetol* 2005;42:104-9.

[52] Lu HL, Wang HW, Wen Y, Zhang MX, Lin HH. Roles of adipocyte-derived hormone adiponectin and resistin in insulin resistance of type 2 diabetes. *World J Gastroenterol* 2006;12:1747-51.

[53] Tokuyama Y, Osawa H, Ishizuka T, Onuma H, Matsui K, Egashira T et al. Serum resistin level is associated with insulin sensitivity in Japanese patients with type 2 diabetes mellitus. *Metabolism* 2007; 56:693-8.

[54] Yin J, Gao H, Yang J, Xu L, Li M. Measurement of salivary resistin level in patients with type 2 diabetes. *Int J Endocrinol* 2012; 2012:359724.

[55] Gharibeh MY, Al Tawallbeh GM, Abboud MM, Radaideh A, Alhader AA, Khabour OF. Correlation of plasma resistin with obesity and insulin resistance in type 2 diabetic patients. *Diabetes Metab* 2010;36:443-9.

[56] Kocot J, Dziemidok P, Kiełczykowska M, Kurzepa J, Szcześniak G, Musik I. Is There Any Relationship between Plasma 25-

Hydroxyvitamin $D_3$, Adipokine Profiles and Excessive Body Weight in Type 2 Diabetic Patients? *Int J Environ Res Public Health* 2018;15:19.

[57] Palanivel R, Maida A, Liu Y, Sweeney G. Regulation of insulin signalling, glucose uptake and metabolism in rat skeletal muscle cells upon prolonged exposure to resistin. *Diabetologia* 2006;49:183-90.

[58] Fan HQ, Gu N, Liu F, Fei L, Pan XQ, Guo M et al. Prolonged exposure to resistin inhibits glucose uptake in rat skeletal muscles. *Acta Pharmacol Sin* 2007;28:410-6.

[59] Rajala MW, Obici S, Scherer PE, Rossetti L. Adipose-derived resistin and gut-derived resistin-like molecule-beta selectively impair insulin action on glucose production. *J Clin Invest* 2003;111:225-30.

[60] Liu F, Yang T, Wang B, Zhang M, Gu N, Qiu J et al. Resistin induces insulin resistance but does not affect glucose output in rat-derived hepatocytes. *Acta Pharmacol Sin* 2008;29:98-104.

[61] Al-Suhaimi EA, Shehzad A. Leptin, resistin and visfatin: the missing link between endocrine, metabolic disorders and immunity. *Eur J Med Res* 2013;18:12.

[62] Palik E, Baranyi E, Melczer Z, Audikovszky M, Szöcs A, Winkler G et al. Elevated serum acylated (biologically active) ghrelin and resistin levels associate with pregnancy-induced weight gain and insulin resistance. *Diabetes Res Clin Pract* 2007;76:351-7.

[63] Mu H, Ohashi R, Yan S, Chai H, Yang H, Lin P et al. Adipokine resistin promotes *in vitro* angiogenesis of human endothelial cells. *Cardiovasc Res* 2006;70:146-57.

[64] Shetty GK, Economides PA, Horton ES, Mantzoros CS, Veves A. Circulating adiponectin and resistin levels in relation to metabolic factors, inflammatory markers, and vascular reactivity in diabetic patients and subjects at risk for diabetes. *Diabetes Care* 2004;27: 2450-7.

[65] On YK, Park HK, Hyon MS, Jeon ES. Serum resistin as a biological marker for coronary artery disease and restenosis in type 2 diabetic patients. *Circ J* 2007;71:868-73.

[66] Tsiotra PC, Tsigos C, Anastasiou E, Yfanti E, Boutati E, Souvatzoglou E et al. Peripheral mononuclear cell resistin mRNA expression is increased in type 2 diabetic women. *Mediators Inflamm* 2008; 2008:892864.

[67] Curat CA, Wegner V, Sengenès C, Miranville A, Tonus C, Busse R et al. Macrophages in human visceral adipose tissue: increased accumulation in obesity and a source of resistin and visfatin. *Diabetologia* 2006;49:744-7.

[68] Al-Daghri NM, Al-Attas OS, Hussain T, Sabico S, Bamakhramah A. Altered levels of adipocytokines in type 2 diabetic cigarette smokers. *Diabetes Res Clin Pract* 2009;83:e37-9.

[69] Burnett MS, Lee CW, Kinnaird TD, Stabile E, Durrani S, Dullum MK et al. The potential role of resistin in atherogenesis. *Atherosclerosis* 2005;182:241-8.

[70] Tuttolomondo A, La Placa S, Di Raimondo D, Bellia C, Caruso A, Lo Sasso B et al. Adiponectin, resistin and IL-6 plasma levels in subjects with diabetic foot and possible correlations with clinical variables and cardiovascular co-morbidity. *Cardiovasc Diabetol* 2010;9:50.

[71] Uslu S, Kebapçi N, Kara M, Bal C. Relationship between adipocytokines and cardiovascular risk factors in patients with type 2 diabetes mellitus. *Exp Ther Med* 2012;4:113-20.

[72] Osawa H, Ochi M, Kato K, Yamauchi J, Nishida W, Takata Y et al. Serum resistin is associated with the severity of microangiopathies in type 2 diabetes. *Biochem Biophys Res Commun*2007;355:342-6.

[73] Verma S, Li SH, Wang CH, Fedak PW, Li RK, Weisel RD et al. Resistin promotes endothelial cell activation: further evidence of adipokine-endothelial interaction. *Circulation* 2003;108:736-40.

[74] Kawanami D, Maemura K, Takeda N, Harada T, Nojiri T, Imai Yet al. Direct reciprocal effects of resistin and adiponectin on vascular endothelial cells: a new insight into adipocytokine-endothelial cell interactions. *Biochim Biophys Res Commun* 2004;314:415-9.

[75] Rizzo M, Rini GB, Berneis K. The clinical relevance of LDL size and subclasses modulation in patients with type-2 diabetes. *Exp Clin Endocrinol Diabetes* 2007;115:477-82.

[76] Zhang MH, Na B, Schiller NB, Whooley MA. Association of resistin with heart failure and mortality in patients with stable coronary heart disease: data from the heart and soul study. *J Card Fail* 2011;17: 24-30.

[77] Muse ED, Feldman DI, Blaha MJ, Dardari ZA, Blumenthal RS, Budoff MJ et al. The association of resistin with cardiovascular disease in the Multi-Ethnic Study of Atherosclerosis. *Atherosclerosis* 2015;239:101-8.

[78] Kreçki R, Krzemińska-Pakuła M, Peruga JZ, Szcześniak P, Lipiec P, Wierzbowska-Drabik K et al. Elevated resistin opposed to adiponectin or angiogenin plasma levels as a strong, independent predictive factor for the occurrence of major adverse cardiac and cerebrovascular events in patients with stable multivessel coronary artery disease over 1-year follow-up. *Med Sci Monit* 2011;17:CR26-32.

[79] Hu WL, Qiao SB, Hou Q, Yuan JS. Plasma resistinis increased in patients with unstable angina. *Chin Med J* (Engl) 2007;120:871-5.

[80] Lim S, Koo BK, Cho SW, Kihara S, Funahashi T, Cho YM et al. Association of adiponectin and resistin with cardiovascular events in Korean patients with type 2 diabetes: the Korean atherosclerosis study (KAS): a 42-month prospective study. *Atherosclerosis* 2008;196: 398-404.

[81] Luc G, Empana JP, Morange P, Juhan-Vague I, Arveiler D, Ferrieres J et al. Adipocytokines and the risk of coronary heart disease in healthy middle-aged men: the PRIME Study. *Int J Obes* (Lond); 2010;34: 118-26.

[82] Fu Y, Yu Y, Wu Y, You Y, Zhang Y, Kou C. Association between Two Resistin Gene Polymorphisms and Metabolic Syndrome in Jilin, Northeast China: A Case-Control Study. *Dis Markers* 2017; 2017:1638769.

[83] Hivert MF, Manning AK, McAteer JB, Dupuis J, Fox CS, Cupples LA et al. Association of variants in RETN with plasma resistin levels and diabetes-related traits in the Framingham Offspring Study. *Diabetes* 2009;58:750-6.

[84] Fontana A, Ortega Moreno L, Lamacchia O, De Bonis C, Salvemini L, De Cosmo S et al. Serum resistin is causally related to mortality risk in patients with type 2 diabetes: preliminary evidences from genetic data. *Sci Rep* 2017;7:61.

[85] Kocot J, Dziemidok P, Kiełczykowska M, Hordyjewska A, Szcześniak G, Musik I. Adipokine Profile in Patients with Type 2 Diabetes Depends on Degree of Obesity. *Med Sci Monit* 2017;23:4995-5004.

[86] Hansen D, Dendale P, Beelen M, Jonkers RA, Mullens A, Corluy L et al. Plasma adipokine and inflammatory marker concentrations are altered in obese, as opposed to non-obese, type 2 diabetes patients. *Eur J Appl Physiol* 2010;109:397-404.

[87] Mohammadzadeh G, Zarghami N, Mobaseri M. Serum Resistin Concentration in Obese Diabetic patients: Any Possible Relation to Insulin Resistance Indices? *Int J Endocrinol Metab* 2008;4:183-193.

[88] Yaturu S, Daberry RP, Rains J, Jain S. Resistin and adiponectin levels in subjects with coronary artery disease and type 2 diabetes. *Cytokine* 2006;34:219-23.

[89] AL-Suhaimi EA, AL-Kulaifi FM, Shehzad A. Serum Adipokines Metabolic Function and Immunological Correlations in Treated Type 2 Diabetes mellitus and Obese Subjects. *World Applied Sciences Journal* 2013;22:933-8.

[90] Al Sheikh MH. The Determinants of Leptin Levels in Diabetic and Nondiabetic Saudi Males. *Int J Endocrinol* 2017;2017:3506871.

[91] Mabrouk R, Ghareeb H, Shehab A, Omar K, El-Kabarity RH, Soliman D et al. Serum visfatin, resistin and IL-18 in A group of Egyptian obese diabetic and non-diabetic individuals. *Egypt J Immunol* 2013;20:1-11.

In: Resistin
Editor: Gerald Maldonado
ISBN: 978-1-53614-543-4


*Chapter 4*

# RESISTIN IN CARDIOVASCULAR DISEASES

***Monica Tucureanu and Ileana Manduteanu****
Department of Biopathology and Therapy of Inflammation
Insitute of Cellular Biology and Pathology, Bucharest, Romania

## ABSTRACT

Resistin is a 12.5 kD cysteine-rich protein belonging to the family of proteins named FIZZ (found in inflammatory zone) or RELM (resistin-like molecules). Resistin has an established function in mice as a molecular link between obesity and diabetes and an unclear function in humans, were it is presumed to be primarily involved in inflammation. In mice, resistin is produced by adipocytes, while in humans it is produced by neutrophils, monocytes, and macrophages. The human resistin gene is homologous by ~60% with mice resistin, explaining the differences in properties of the protein between rodents and humans. Resistin in humans is known to be expressed in atherosclerotic plaque, and in aneurisms and its plasma levels are associated with coronary heart disease (CHD). Accumulating data suggest that resistin may be involved in the initiation, progression and rupture of atherosclerotic plaque. Recent data reported that resistin may promote the initiation of atherogenesis by activating vascular endothelial cells (EC) and smooth muscle cells (SMC) and by interfering in their cross-

* Corresponding Author Email: ileana.manduteanu@icbp.ro.

talk, leading to an increase in vessel wall inflammation. Resistin, when present in the diseased vessel wall, may orchestrate a cascade of pro-atherosclerotic events, contributing to atherosclerosis and progression towards plaque rupture. Thus, resistin may augment monocyte and macrophage infiltration, may interfere in cell cross-talk, and may modulate the expression of various inflammatory cytokines, leading to acceleration of plaque growth and destabilization by mechanisms that are now being uncovered. Existence of a specific resistin receptor in immune and vascular cells is controversial, but recent data indicating that the Toll-like receptor 4 (TLR4) and Gi protein signalling are involved in resistin effects in vascular cells and that adenylyl cyclase-associated protein 1 (CAP1) may function as a receptor for human resistin. Resistin and fractalkine, which is an atypical chemokine that functions as chemoattractant or as adhesion molecule, were found at increased levels in human atherosclerotic plaque, in the sub-endothelium, resistin being associated with CD68 monocytes/macrophages and fractalkine predominantly associated with the intimal SMC and with monocytes/macrophages. Recent data uncovered fractalkine as a new molecular link between resistin and inflammation. This chapter reviews resistin's role in vascular inflammation in atherosclerosis, the molecules and mechanisms involved, and their potential use as new targets for innovating nanobiotherapies.

## 1. INTRODUCTION

Acute and chronic inflammation associated with various pathologies (i.e., obesity, diabetes, rheumatoid arthritis) have been reported to be implicated in the onset, pathogenesis, and progression of cardiovascular diseases (CVD); but the molecular links still need to be defined. In this context, cytokines and chemokines may be important players. Resistin is a 12.5 kDa cysteine-rich polypeptide; is a member of a family of proteins named FIZZ (found in inflammatory zones) or RELM (resistin-like molecules); and belongs to a group of adipose tissue-related hormones (termed adipokines) that includes leptin, adiponectin, adipsin, and visfatin (Gualillo, González-Juanatey, and Lago, 2007; Lee and Kim, 2012; Schwartz and Lazar, 2011).

The term "resistin" was first introduced in 2001 by Steppan et al. (Steppan et al., 2001) to reveal a murine protein secreted by adipocytes as a hormone that potentially links obesity to diabetes. But extensive research demonstrated the complex role of resistin and the differences between mouse and human regarding its function and mechanisms of action in various inflammatory pathologies. Human resistin shares only 55% amino acid homology with that of rodents, which may explain the different properties of the protein in rodents vs. humans (Park and Ahima, 2013; Park et al., 2017). In mice, resistin is well known as a molecular link between obesity and diabetes. In humans, where it is produced by neutrophils, monocytes and macrophages (Kunnari et al., 2009), resistin is mainly reported as a pro-inflammatory molecule associated with both acute and chronic inflammation and its specific functions are yet to be uncovered (Huang and Yang, 2016; Nagaev et al., 2006).

Numerous studies have demonstrated that secreted resistin may have pleiotropic effects, through endocrine, paracrine and autocrine mechanisms. Many cell types and tissues respond to resistin action, involving this protein in multiple physiological and pathological processes (Barnes and Miner, 2009; Jamaluddin et al., 2012; Pine, Batugedara and Nair, 2018). Accumulating epidemiological, genetic and clinical studies indicate that human resistin is involved in the pathogenesis of inflammation, insulin resistance, diabetes and cardiovascular diseases (CVD). It is assumed that resistin plays an important role in cardiovascular disease by interfering in systemic and local inflammation in the diseased vessel wall (Kawanami et al., 2004; Reilly et al., 2005; Verma et al., 2003).

Resistin was detected at the atherosclerotic plaque level and in human aneurisms associated with CD68 positive macrophages (Burnett et al., 2005; Jung et al., 2006). Recent *in vivo* studies indicated resistin as an important contributor to atherosclerosis by augmenting monocyte and macrophage infiltration and expression of various inflammatory cytokines, resulting in the acceleration of plaque growth and destabilization (Cho et al., 2011). It was also shown that resistin induced vascular wall cells inflammation (Calabro et al., 2004; Verma et al., 2003) and accelerated foam cell formation (Xu et al., 2006).

Controversies concern the existence of a specific resistin receptor in immune and vascular cells, but recent data indicate that Toll-like receptor 4 (TLR4) and Gi protein signalling are involved in resistin effects in vascular cells (Pirvulescu et al., 2014; Tarkowski et al., 2010) and that adenylyl cyclase-associated protein 1 (CAP1) may function as a receptor for human resistin (Lee et al., 2014). Further work is necessary to identify the specific receptor for human resistin, which would lead to delineation of specific signalling pathways and aid in the clarification of the role of resistin in the pathogenesis and treatments of type 2 diabetus mellitus (DM), dyslipidemia and CVD.

This chapter reviews resistin effects and possible operating mechanisms in CVD. We focus on resistin effects and mechanisms in EC, (SMC), monocytes/macrophages, as well as subendothelial resistin (sR) effects and mechanisms in EC, SMC, EC-SMC cross-talk, SMC-monocytes/ macrophages cross-talk and in the subsequent monocyte transmigration and macrophage accumulation in the diseased vessel wall. Moreover, we review the possible exploitation of resistin-dependent mechanisms as targets for innovative therapies.

## 2. Resistin Detection in CVD Patients' Plasma, in the Inflamed Vessel Wall and in Atheroma

Clinical studies have demonstrated the role of resistin in CVD. An increased level of plasma resistin was positively correlated with atherosclerotic markers and lesion progression, with a link between obesity and cardio-metabolic risk (Ding et al., 2011). Moreover, plasma resistin was associated with coronary atherosclerotic lesion and ischemic stroke (Osawa et al., 2009). In addition, accumulating clinical data indicate resistin as an independent predictor of major adverse cardiovascular events, including restenosis, myocardial infarction and death in patients undergoing coronary interventions (On et al., 2007). A Multi-Ethnic Study of Atherosclerosis (MESA) assessing the relationship between circulating resistin levels and cardiovascular outcomes over a seven-year period found a strong, independent association between higher resistin levels and incident CVD, coronary heart disease (CHD) and heart failure (HF), independent of traditional cardiovascular risk factors, obesity, markers of inflammation/ insulin resistance and several adipokines (Muse et al., 2015). Increased serum resistin levels were associated with exacerbated coronary artery disease (CAD) in patients with carotid artery stenosis (Wang et al., 2009). Resistin was found to be elevated in obese and diabetic individuals (Vozarova de Courten et al., 2004; Wolk et al., 2004) and has been associated with CVD and poor clinical outcomes (Owens et al., 2010). A meta-analysis of all published prospective studies (at the 2015 level) on the association between circulating resistin and mortality risk of either all-cause or cardiovascular origin, provided evidence for an association between circulating resistin and mortality risk among high-risk patients as are those with diabetes and CAD (Fontana et al., 2015). A recent study showed that resistin levels were higher in patients with CAD with high aortic stiffness than in those with low aortic stiffness; and together with DM, older age and hs-CRP, resistin level was an independent predictor for the development of aortic stiffness in these patients. In addition, DM, age and waist circumference positively correlated with resistin levels in patients with CVD (Wang et al., 2017).

Circulating resistin levels were correlated with inflammatory and fibrinolytic markers, such as C-reactive protein (CRP), TNF-α and IL-6 in in the general population and in individuals with DM, coronary atherosclerosis, chronic kidney disease, rheumatoid arthritis and/or sepsis (Park and Ahima, 2013).

Resistin levels were also found to be associated with plasma CRP levels in the Study of Inherited Risk of Coronary Atherosclerosis (SIRCA) (Reilly et al., 2005). Moreover, recent data suggested that in humans, a multi-cytokine resistin pathway might operate, a pathway that was associated with cardiovascular risk factors and which strongly and independently predicted major adverse cardiovascular events (Menzaghi et al., 2017). Concerning the cardiovascular risk assessment, diagnosis and prognosis, resistin was revealed as a predictor of HF risk in non-CHD community dwelling-individuals (Frankel et al., 2009).

A recent study involving over 250 elderly patients with T2DM and mild, moderate or severe carotid atherosclerosis (characterized by ultrasonic examination) showed that the circulating levels of resistin, vaspin and visfatin gradually increased with the severity of atherosclerosis and were positively correlated with body mass index (BMI) and a homeostasis model assessment of insulin resistance (HOMA-IR), concluding that adipose tissue-derived inflammatory factors may be involved in the pathogenesis of atherosclerosis in T2DM patients (Yang et al., 2018).

A recent literature search of potential cardiovascular biomarkers regulating atherosclerosis, inflammation, endothelial dysfunction and fibrosis in cardiovascular disease revealed, by reviewing over 900 papers, that resistin may be a predictor of CVD, disease severity and outcome and prognosis of CVD (Hogas et al., 2017). Resistin was proposed to be an important player in cardiovascular dysfunction and was proposed as a biomarker of endothelial dysfunction, atherosclerosis, inflammation, fibrosis and atherothrombosis (Hogas et al., 2017).

In prospective studies, high resistin levels significantly increased the risk for all CVD, chronic CHD (Muse et al., 2015) and acute coronary syndrome, independently of other cardiovascular risk factors, including the inflammatory markers IL-6 and CRP (Weikert et al., 2008). In patients with multivessel coronary disease, a resistin concentration over 17.3 ng/ml was associated with a 13-fold increase in the risk of major cardiac and cerebrovascular events (Kreçki et al., 2011).

Resistin was found at increased levels in atherosclerotic regions of the human carotid artery and aorta, but not in the normal vessel wall, indicating that it may represent a biomarker and a contributor of atherosclerosis evolution (Burnett et al., 2005; Chen et al., 2010).

However, in the prospective study of Zhang et al. (Zhang et al., 2011), resistin did not provide prognostic information beyond traditional risk factors regarding the development of HF in CHD patients. Therefore, resistin may be proposed as a biomarker for HF in non-CHD individuals.

Several studies have suggested that resistin may be proposed as a biomarker for HF risk assessment and stratification (Butler et al., 2009;

Frankel et al., 2009; Yanavitski and Givertz, 2011). Resistin was also proposed as a biomarker for acute myocardial infarction, since resistin levels were higher in AMI than in unstable/stable angina patients (Qiao et al., 2007). Resistin increased 3-6 hours after chest pain onset and peaked 12 hours after chest pain onset (Lubos et al., 2007). However, studies comparing resistin to other classical biomarkers are needed (Hogas et al., 2017).

## 3. Atherosclerosis Stages and Potential Resistin Involvement

A main cause of CVD is atherosclerosis, the most common trigger of heart attacks, strokes and peripheral vascular diseases. Moreover, atherosclerosis is accelerated in diabetes. In atherosclerosis, immune cells, vascular cells and different effector molecules cross-talk and determine the speed and evolution of the disease by specific mechanisms that remain to be clarified and may help to reveal new targets for prevention or cure of the diseases. Based on our data and from the literature, in order to better understand the pathophysiology of the disease, based on the importance of the phenomena from a developmental point of view, we proposed that although atherosclerosis is a continuous process, a set of consecutive stages delineate atheroma development and fate; we provided an overview on the intersection of inflammation and dyslipidemia in atherosclerosis (Manduteanu and Simionescu, 2012).

The concept of inflammation in atherosclerosis does not diminish the importance of traditional risk factors for atherosclerosis. The concept of inflammatory signalling and the participation of proinflammatory cytokines offer a molecular link between the traditional risk factors and the altered biological responses in the arterial wall, which determines atherosclerosis and its complications.

## 3.1. Atherosclerosis—a Continuous Process with Consecutive Stages

Atheroma development is a gradual and continuing process. But the progressive implication of vascular resident and nonresident cells and their secretory products may define an arbitrary sequence of consecutive stages: from fatty streak to fibro-lipid plaque and ultimately to plaque rupture and atherothrombosis. Inflammation is implicated in all stages of atherosclerotic lesion evolution as reviewed in Manduteanu and Simionescu, 2012. Therefore, in stage I, atheroma formation is initiated by EC activation/modulation of constitutive functions, leading to stage II, EC dysfunction. Affected on both sides, EC initiates an inflammatory process manifested by the expression of new or more cell adhesion molecules, cytokines and chemokines, an indication ofendothelial dysfunction. In stage III, recruitment of blood immune cells and commencement of a robust inflammatory reaction occur. Blood monocytes and T cells adhere to activated/dysfunctional EC and are recruited into the intima. Adhered platelets assist leucocyte migration. Within the intima, monocytes become activated macrophages and foam cells are formed and secrete a variety of proinflammatory (Th1) mediators. Lymphocytes shift to activated Th1 and anti-inflammatory (Th2 and Treg) cells that secrete cytokines and chemokines. The direct or indirect cross-talk between resident and migrated cells within the intima dictates the lesion progression to stage IV, where SMCs are key participants in fibrous plaque formation, and stage V, where resident and immune cells and the factors they secrete generate a calcified fibro-lipid plaque, which progresses toward stage VI, the unstable fibro-lipid plaque, and ultimately generates the physical rupture of the plaque. This ends in direct contact between tissue factors and circulating platelets and blood coagulation components triggering the thrombus formation that may partially or totally impede the blood flow leading to either myocardial infarction or stroke (Manduteanu and Simionescu, 2012).

Since resistin was associated with atherosclerosis and inflammation (Reilly et al., 2005), particularly with early stages with EC activation/ dysfunction (Yang et al., 2012) and with the late stages of atheroma vulnerability and rupture (Gasbarrino et al., 2016), we may presume an important role both in early atherosclerosis and in the evolution towards vulnerable plaque and rupture. We will review important findings that may help to advance the understanding of resistin-dependent mechanisms involved.

## 4. Resistin's Role and Mechanisms in Initiation of Atherosclerosis

### 4.1. Resistin Activation of Endothelial Cells

Recent data reported that resistin may promote the initiation of atherogenesis by activating vascular EC and SMC and by interfering in their cross-talk, leading to an increase in monocyte infiltration and vessel wall inflammation (Codoñer-Franch and Alonso-Iglesias, 2015; Pirvulescu et al., 2014).

Endothelial dysfunction consists of, among other processes, loss of NO bioavailability, that, together with an increased expression of vasoconstriction-associated molecules, leads to vascular homeostasis alteration. Recent studies have shown that resistin decreased the expression of NO in HEC by increasing ROS levels and reducing NOS phosphorylation (Chen et al., 2010; Ou et al., 2012). Thus one can assume that the balance between oxidative stress and resistin may be an early event and a central link between obesity, inflammation and atherosclerosis.

In EC, human resistin was shown to increase the expressions of various proinflammatory factors, such as MCP-1, fractalkine, endothelin-1 and matrix metalloproteinases, as well as adhesion molecules including ICAM-1, VCAM-1, PECAM-1 and P-selectin (Hsu et al., 2011; Kawanami et al., 2004; Reilly et al., 2005; Verma et al., 2003). We have reported that resistin up-regulated the gene and protein expression of P-selectin and fractalkine in HEC by a mechanism involving ROS, MAPK and the transcription factors NF-kB and AP-1 (Manduteanu et al., 2009, 2010). Evidence including our data indicates that resistin may also interfere in cytokine signalling in HEC (Figure 1) and could be a significant modulator of inflammation, acting as a function of the specific inflammatory microenvironment (Manduteanu et al., 2009; Verma et al., 2003).

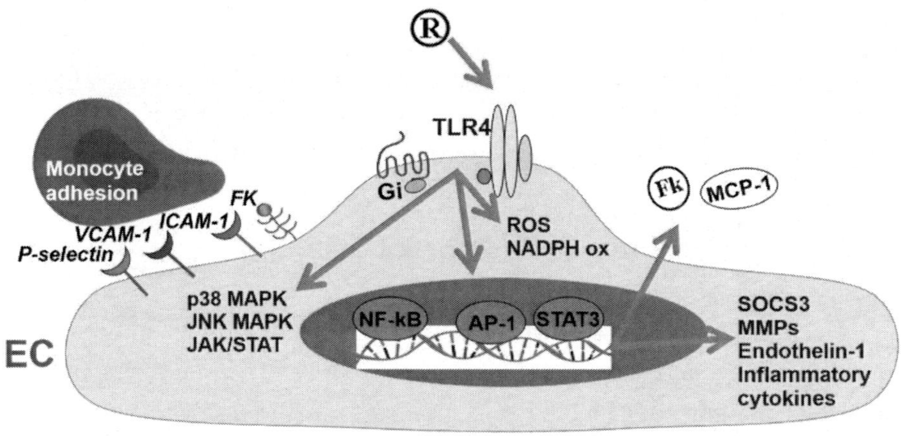


Figure 1. Resistin (R) proinflammatory effects in endothelial cells (EC). Resistin activates TLR4 and Gi protein signalling and amplifies the expression of cellular adhesion molecules (CAM), such as VCAM-1, ICAM-1, P-selectin and fractalkine (Fk). The mechanisms include increases in reactive oxygen species (ROS) levels and activation of the NAPDH oxidase (NADPH ox), p38 and JNK MAPK and JAK/STAT pathway. Moreover, resistin activates NF-kB, AP-1 and STAT3 transcription factors and increases endothelin-1, SOCS3, MMPs and inflammatory cytokines expression. Resistin also enhances the secretion of FK and MCP-1 chemokines, that, together with CAMs, may be instrumental in monocyte adhesion and transmigration.

Human resistin also increased endothelial permeability through mechanisms involving oxidative stress and the activation of p38 MAPK, which may promote endothelial cell/monocyte adhesion and infiltration (Jamaluddin et al., 2013). These resistin actions were mediated by the ERK and p38 mitogen-activated protein kinase signalling pathways (Hsu et al., 2011). Furthermore, resistin inhibited endothelial nitric oxide synthase via oxidative stress in human endothelial cells (Chen et al., 2010).

Fractalkine is an adhesion molecule, a soluble chemotactic factor mediating firm adhesion and transmigration of leukocytes expressing cognate receptors, such as monocytes, cytotoxic T lymphocytes or natural killer cells (Umehara et al., 2004). Fractalkine is expressed by different cell types, such as endothelial cells, vascular SMC, dendritic cells and neurons (Ahn et al., 2004; Fraticelli et al., 2001). Leukocyte adhesion mediated by fractalkine was shown to play important roles in mediating inflammatory diseases such as CAD (Damås et al., 2005), arthritis, systemic lupus erythematosus (Ruth et al., 2001; Yajima et al., 2005) and brain inflammation (Chapman et al., 2000). Fractalkine expression was detected in human atherosclerotic plaques (Stolla et al., 2012) and on inflamed or atherosclerotic endothelial cells and was shown to mediate direct capture and firm adhesion of circulating leukocytes under conditions of low shear forces and high arterial shear stress (Schulz et al., 2007). In recent studies fractalkine was detected in all stages of human atherosclerotic plaque formation and the number of fractalkine-expressing cells positively correlates with the number of CX3CR1-positive cells in human carotid artery plaques (Stolla et al., 2012). Fractalkine was shown to be expressed at low levels in normal endothelial cells but was markedly increased by proinflammatory agents like TNF-α, LPS, IL-1, IFN-γ and CD40L; while others, such as IL-4 and IL-13, inhibit its expression (Fraticelli et al., 2001). Moreover, it is considered that highly elevated levels of fractalkine are positively correlated with massive monocyte activation and accelerated atherosclerotic plaque progression, leading to plaque rupture and myocardial infarction (Skoda, Stangret and Szukiewicz, 2018).

We have shown that resistin induced fractalkine in HEC by mechanisms involving NF-kB, p38 and JNK MAPK (Manduteanu et al., 2009). Our data showed, similar to previous studies, that resistin activates directly the p38 and JNK MAPK signalling. Moreover, we compared resistin effects with that of the prototype inflammatory cytokine, TNF-α. Since resistin and TNF-α are often found together in the inflammatory areas, we also evaluated the effect of co-stimulation of EC with both activators on fractalkine induction. Interestingly, we found that resistin induced fractalkine expression as with TNF-α. When acting together, the mediators did not have additive effects (Manduteanu et al., 2009) on fractalkine induction or on the activation of signalling proteins involved in fractalkine up-regulation, suggesting that the simultaneous presence of resistin and TNF-α does not necessarily lead to an increase in the inflammatory process. Since in EC, TNF-α signalling events are dependent on the interaction of TNF-α with TNFR1 (Luo et al., 2006), we assessed whether resistin influenced receptor expression in HEC. We found that the constitutive expression of TNFR1 was not significantly changed by resistin; while in the case of co-stimulation of HEC with resistin and TNF-α, receptor expression was reduced to a value lower than the level induced by TNF-α acting alone. Thus we may presume that when both cytokines are present, resistin might interfere with the signalling pathway of TNF-α. In contrast, it has been reported that resistin had additive effects with CD40L in MCP-1 production (Verma et al., 2003) without altering CD40 receptor expression. Together, these data and our results may indicate that resistin is an important modulator of inflammation, acting as a function of a specific inflammatory microenvironment. Our study added new data on the inflammatory potential of resistin, shedded new light on the cross-talk between resistin and TNF-α in HEC and uncovered fractalkine as a new molecular link between resistin and inflammation (Manduteanu et al., 2009). However, the role of resistin in the up-regulation of fractalkine, ICAM-1, VCAM-1 and MCP-1 (Kawanami et al., 2004; Verma et al., 2003) indicates its potential contribution in the initiation and evolution of the inflammatory process. Since it was shown that concomitant expression of fractalkine, ICAM-1 and VCAM-1 in enhancement of leukocyte adhesion was comparable with each system alone (Schulz et al., 2007), we may suppose that resistin-induced fractalkine expression might contribute to the enhancement of leukocyte adhesion and increase the inflammatory process.

#### *4.1.1. Resistin Up-Regulated SOCS3 Expression in EC through STAT3 Activation*

We and others have shown that resistin-activated endothelial cells (inducing the up-regulation of cell adhesion molecules) interfere in cytokine signalling and could contribute to tailoring of the micro-inflammatory milieu (Manduteanu et al., 2009; Verma et al., 2003). Considering that resistin is a proinflammatory molecule which may interfere with cytokine signalling in HEC and may interfere with up-regulated suppressor of cytokine signalling (SOCS3) in mice adipocytes, we examined the effect of resistin on the expression of SOCS3 proteins, molecules that are induced in a cytokine and cell specific manner. Suppressors of cytokine signalling (SOCS) are a family of cytoplasmic proteins that carry out a negative feedback loop to attenuate signal transduction induced by cytokines or hormones such as leptin and insulin (Bjørbaek et al., 1999). Signal transducer and activator of transcription (STAT) proteins are involved in SOCS expression in various cell types, including endothelial cells (O'Sullivan et al., 2007). It has been reported that in mouse 3T3-L1 adipocytes, resistin markedly increased the expression of SOCS3 proteins, which mediated resistin antagonism to insulin; and that loss of SOCS3 function impairs resistin action (Steppan et al., 2005). There are no data on resistin effects on SOCS3 expression in other vascular cells, but there is evidence that resistin induces up-regulation of SOCS3 and inhibits JAK2/STAT3 signalling (essential for cancer cell proliferation) in colon cancer cells, causing the arrest of cells in G1 phase (Singh et al., 2017).

In a previous study we hypothesized that resistin action on HEC may also implicate SOCS3 protein and we searched for the signalling pathways implicated. The experiments showed: (1) exposure to resistin up-regulated SOCS3 mRNA and protein expression in HEC; (2) increased SOCS3 mRNA and protein in resistin-activated HEC was significantly augmented starting with two hours of cell stimulation and the effect persisted up to 18 hr.; (3) activation of the STAT3 signalling pathway was implicated in the process; (4) STAT3 activation was significant after 1 hr. of cell stimulation and continued for 18 hr.; (5) silencing of SOCS3 impeded resistin-induced expression of endothelial cell adhesion molecules in endothelial cells (Pirvulescu et al., 2012). It was demonstrated in endothelial cells that SOCS3 was induced by IL-6 and IFN gamma with a maximum at 2 hr. for mRNA and 4 hr. for the proteins, after which they rapidly decreased to the baselines levels (Bluyssen et al., 2010). Interestingly, our data showed that, like IL-6/IFN gamma, resistin induced up-regulation of SOCS3 expression in HEC at 2 hr., but in contrast, the SOCS3 gene and protein expression was prolonged for up to 18 hr. We supposed that this may indicate either a special feature of resistin signalling in endothelial cells and/or involvement of other inflammatory mediators produced by resistin in these cells. It has been shown that resistin has stimulated IL-6 expression in mouse endothelial cells (Cho et al., 2011). Since IL-6 induces SOCS3 expression (Bluyssen et al., 2010), we presumed that the combined effect of these cytokines might contribute to extended expression of SOCS3 in HEC. But further experiments are needed to clarify this matter.

Since it was shown that SOCS3 is a target gene for STAT3 transcription factor in endothelial cells (Ortiz-Munoz et al., 2009), we assessed whether resistin activated the STAT3 pathway. Our data showed that after 1 hr. of stimulation, resistin activated the STAT3 transcription factor and the effect lasted longer (up to 18 hr.) compared to the action of IL-6 or IFN-gamma (2-4 hr. respectively) (Bluyssen et al., 2010). These data may indicate the possibility of a complex effect of resistin in the process. In addition, our results indicated that the STAT3 pathway may represent a central regulator of SOCS3 expression in resistin-activated HEC.

In contrast, in studies on mice adipocytes, no increase in phosphorylation of STATs in response to resistin was detected (Steppan et al., 2005). The variance between our results and those obtained on mice adipocytes might be due to species or cell type specific effects of resistin. How resistin activated the STAT3 signalling pathway in HEC and whether the effects of resistin are mediated by one or several receptors in vascular cells still need to be investigated.

We have data indicating that TLR4 may be involved in resistin-induced SOCS3 expression (data not published). We supposed that resistin-induced STAT3 phosphorylation in HEC was due to a direct effect on JAK/STAT pathway activation or acted through TLR4-associated adaptor proteins or through cross-talk among receptors.

Interestingly, our data indicate that SOCS3 expression and activated STAT3 coexisted up to 18 hr. of stimulation of endothelial cells with resistin. In the case of other cytokines, such as IFN-gamma and IL-6, the increase in the SOCS3 protein was accompanied by a reduction of STAT3 phosphorylation, suggesting a regulatory role for SOCS3 in STAT3 signalling in endothelial cells (Bluyssen et al., 2010). The disagreement between these data and our results might be explained by variances in cytokine action (IFN-gamma, IL-6 compared to resistin) and by the possibility that the level of SOCS3 induced by resistin was not high enough to completely shut off STAT3 activation. As a result, SOCS3 and activated STAT3 coexisted for a prolonged time period and could prolong and enhance the inflammatory process in endothelial cells. This supposition was sustained by our results showing that blocking SOCS3 before stimulation with resistin reduced endothelial cell adhesion molecules (fractalkine and P-selectin) expression. The data corroborate well with the reported incomplete suppression of STAT3 by SOCS3 in patients affected by chronic inflammatory diseases, such as rheumatoid arthritis, in which resistin is recognized as an important contributor in these pathologies (Liu et al., 2005; Migita et al., 2006). As such, the extended coexistence of SOCS3 and activated STAT3 induced by resistin may have a role in the perpetuation of the inflammatory process. These results indicate the notable role of SOCS3 in maintaining the proinflammatory state of endothelium and designated SOCS3 as a novel therapeutic target to modulate resistin-dependent inflammation in vessel wall diseases (Pirvulescu et al., 2012).

#### *4.1.2. Resistin and High Glucose Effects and Mechanisms in Endothelial Cells*

High glucose (HG) and increased circulating resistin are concurrently found in the plasma of diabetic patients (Osawa et al., 2009). Since vascular endothelium is directly exposed to any change in plasma homeostasis, our group investigated whether resistin and/or HG contributed to HEC dysfunction by inducing modifications like the expression of P-selectin and fractalkine, adhesion molecules with key roles in leukocyte adhesion and transmigration across endothelium in inflammation.

Our data showed that resistin, high glucose and combined resistin with high glucose up-regulated P-selectin gene and protein expression at similar levels in HEC (Manduteanu et al., 2010). These results contrasted with data from a previous report, where it was shown that resistin did not induce P-selectin in EC (Kunnari et al., 2009); the discrepancy may be due to the lower dose of resistin used, 50 ng/ml versus 100 ng/ml in our study. We considered that the concentration employed in our experiments was clinically relevant, since resistin was reported to induce endothelial dysfunction at concentrations higher than 50 ng/ml (Jung et al., 2006; Verma et al., 2003) to be nontoxic for cultured endothelial cells up to 1000 ng/ml resistin and to induce more adhesion molecules than the lower physiologic plasma concentrations. Furthermore, in response to cytokine treatment, plasma resistin concentration increased by four to five-fold (Lehrke et al., 2004) and a high level of resistin was reportedly present in areas of atherosclerotic lesions (Burnett et al., 2005). Thus in pathological conditions, the vascular endothelium may be exposed to high resistin concentrations. Remarkably, resistin, HG or resistin with HG up-regulated at similar levels the P-selectin expression in HEC, suggesting a similar molecular mechanism. As it was previously shown that HG and resistin have pro-oxidant capacities and that P-selectin expression in EC is oxidant-sensitive (Kashiwagi et al., 1996; Manduteanu et al., 2007), we presumed that the oxidative stress might represent a common mechanism in HG and resistin-induced P-selectin up-regulation in HEC. This hypothesis was supported by our data that resistin, HG or resistin combined with HG increased significantly NADPH oxidase activity and the resulting intracellular ROS in HEC. Moreover, there is evidence that antioxidants, like pyrrolidine dithiocarbamate (PDTC), significantly inhibit cytokine-induced P-selectin expression in human endothelium through an NF-kB independent pathway (Xia et al., 1998). Our data indicate that the antioxidant PDTC inhibited resistin-induced P-selectin mRNA up-regulation in HEC, but additional experiments are needed to elucidate the pro-oxidant mechanisms involved in resistin and HG-induced oxidative stress in HEC (Manduteanu et al., 2010).

Since we have previously shown that resistin up-regulated the fractalkine expression in HEC (Manduteanu et al., 2009), we also evaluated the effect of combined resistin with HG or HG alone on fractalkine induction. Our experiments showed that exposure of HEC to HG resulted in increased fractalkine expression. Co-stimulation of cells with resistin and HG increased the fractalkine expression at statistically similar levels (Manduteanu et al., 2010).

In addition, our data showed that monocyte adhesion was increased in HEC stimulated by resistin, high glucose or combined resistin with high glucose, and demonstrated that fractalkine and P-selectin were functional in the process (Manduteanu et al., 2010).

Partially uncovering the mechanisms involved in the effects of resistin and HG in HEC, we found that the combined inducers provoked an oxidative stress, as documented by the increase in NADPH oxidase activity and ensuing intracellular ROS production. and had an effect similar to that of each inducer alone. Moreover, both inductors activated NF-kB and AP-1 transcription factors at the same level as each inductor alone (Manduteanu et al., 2010); these transcription factors were previously suggested to be implicated in resistin or HG effects (Altannavch et al., 2004; Dragomir et al., 2006; Kawanami et al., 2004; Manduteanu et al., 2009). Since resistin and HG increased intracellular ROS, and it has been shown that the pro-oxidant transcription factors are involved in fractalkine up-regulation in endothelial cells, we presumed that the oxidative stress may represent a common mechanism in fractalkine induction in HEC and that alone or together with other factors may be instrumental in resistin and HG-induced up-regulation of fractalkine expression. Since it has reported that NF-kB and AP-1 are involved in endothelial fractalkine induction by different inflammatory mediators (Manduteanu et al., 2009), we assumed that the lack of an additive effect of resistin and HG on fractalkine up-regulation in HEC might be explained in part by the competition or an ineffective additive effect of the inducers on these transcription factors. In addition, our results have shown that resistin and HG-induced P-selectin and fractalkine were functional in increasing monocyte adhesion to HEC. These data extend previous reports that resistin or HG enhanced monocyte adhesion to endothelial cells (Kunnari et al., 2009; Manduteanu et al., 1999), demonstrating that monocyte adhesion is due to up-regulation of fractalkine and P-selectin, which are known to function in leukocyte recruitment to activated endothelium (Schulz et al., 2007).

Our data might indicate that when present together, resistin and HG did have proinflammatory effects but do not amplify HEC response over that produced by every single inducer, in terms of expression of fractalkine and P-selectin, the level of intracellular ROS, NADPH oxidase activity or transcription factors activation. Other cytokines or inflammatory mediators

(TNFα or LPS) augmented HG effects on HEC (Kinoshita et al., 2006; Manduteanu et al., 1999); these data may suggest that in HG conditions, various mediators may have different effects, the outcome of which may depend on the activated signalling pathways. Our findings uncovered molecular mechanisms involved in the inflammatory effects of resistin and HG on HEC and reveal resistin, fractalkine and P-selectin as possible molecular targets for conceiving antiinflammatory drugs in atherosclerosis and/or diabetes (Manduteanu et al., 2010).

## 4.2. Resistin's Pro-Inflammatory Effects in Smooth Muscle Cells

Smooth muscle cells (SMC) have important roles in early and late atherogenesis. In atherogenesis, a complex interplay between cells and mediators influence plaque evolution. In the process, SMCs switch from a "contractile" to a "synthetic" phenotype and contribute decisively to plaque progression and rupture and inflammatory mediators such as chemokines and cytokines (i.e., TNFα, INFγ) and high glucose have important contributions to this change (Doran, Meller and McNamara, 2008; Manduteanu and Simionescu, 2012). Functional roles of SMC in the pathology of atherosclerosis are described (Alexander and Owens, 2012), involving essentially a phenotypic switching from quiescent "contractile" phenotype to active "synthetic" state, leading to extracellular matrix deposition, proliferation, migration, inflammatory gene expression, oxidant stress and monocyte retention (Figure 2).

Resistin and fractalkine were found at increased levels in human atherosclerotic plaques, fractalkine and its receptor, CX3CR1, being predominantly associated with intimal SMCs. Fractalkine (CX3CL1) is an atypical chemokine that functions both as chemoattractant and as adhesion molecule, facilitating monocyte and T cell transmigration in atherosclerotic lesion prone areas (Fong et al., 1998). As for other chemokines, CX3CL1 recognizes a pertussis toxin-sensitive G-protein coupled receptor, CX3CR1, signalling primarily via the G protein pathway (Al-Aoukaty et al., 1998). CX3CL1 induces mobilization of intracellular calcium and activates the heterotrimeric G proteins, which mediate both leukocyte migration and adhesion (Al-Aoukaty et al., 1998; Imai et al., 1997). CX3CL1 and its cognate receptor are present in human coronary artery atheroma but not in normal arteries, and CX3CR1-positive cells with the characteristics of SMC were present adjacent to fractalkine-positive cells (Lucas et al., 2003).

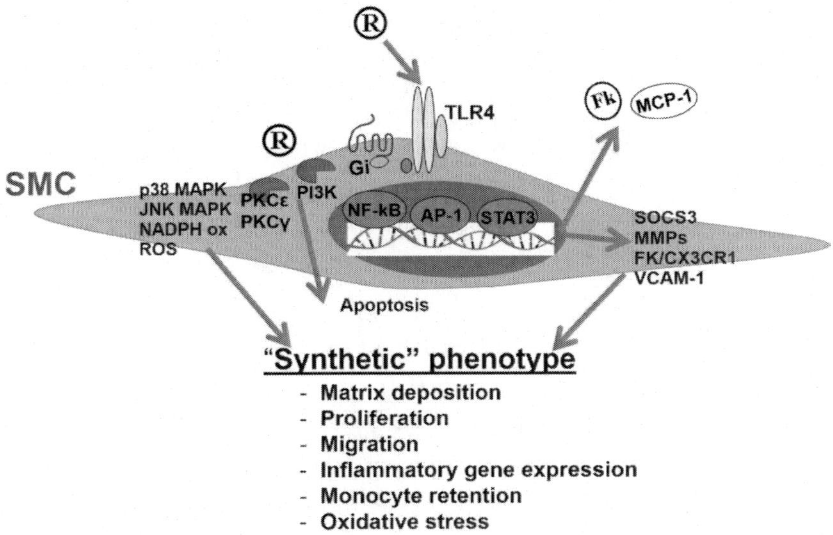


Figure 2. Resistin (R) effects in SMC. In SMC, resistin activates TLR4 and Gi protein signalling, inducing oxidative stress and activation of p38 and JNK MAPK signalling pathways and expression of SOCS3, MMPs, FK/CX3CR1 and VCAM-1. Moreover, resistin induces activation of PI3K, PKCε and PKCγ, leading to cellular apoptosis. By activating NF-kB, AP-1 and STAT3 transcription factors, resistin induces the expression of FK and MCP-1, chemokines with a major role in monocyte transmigration. All these mechanisms cause a phenotypic switch from "contractile" to a "synthetic" state, leading to extracellular matrix deposition, proliferation, migration, inflammatory gene expression, oxidative stress and monocyte retention.

Human SMCs express the CX3CR1 and undergo specific receptor-mediated chemotaxis to CX3CL1 *in vitro* (Lucas et al., 2003). The CX3CL1/CX3CR1 axis is implicated in the pathogenesis of vascular dysfunction. Recently we demonstrated that cross-talk between monocytes and SMC amplifies the inflammatory response in both cell types via the CX3CL1/CX3CR1 axis that functions as inductor of critical molecules for atheroma progression (Butoi, Dragomir et al., 2011). Interrupting the

CX3CL1–CX3CR1 binding *in vivo* has a highly protective effect in animal models of atherosclerosis (Lesnik, Haskell and Charo, 2003). Human subjects heterozygous for the CX3CR1-M280 allele, which encodes a defective receptor, have a reduced risk of atherosclerosis (McDermott et al., 2003). Since both, resistin (a cytokine) and CX3CL1 (a chemokine) are present in the atherosclerotic lesion and the latter and its receptor are predominantly associated with SMC, we hypothesized that resistin may be an inducer of the expression of CX3CL1 and its receptor, thus contributing to the proinflammatory switch of SMC and enhanced monocyte transmigration. As for the effect of resistin on vascular SMC, it was only reported that it induced cell proliferation through activation of ERK1/2 and phosphatidylinositol 3-kinase (Calabro et al., 2004) and up-regulated VCAM-1 expression (Cho et al., 2011).

Our recent results identified resistin as a novel inducer of the proinflammatory phenotype of human SMC and showed that resistin initiated the up-regulation of CX3CL1 and CX3CR1 expression by activating p38 MAPK and JAK/STAT pathways and NF-kB, AP-1 and STAT1/3 transcription factors. In addition, we have found that TLR4 and Gi-protein signalling were operational in the resistin effect on SMC, and STAT1/3 transcription factors are involved in the up-regulation of CX3CL1 and CX3CR1. We have also shown that resistin-induced CX3CL1 is functional in monocyte chemotaxis. These data extend our knowledge of the proinflammatory role of resistin on vascular cells and reveal the potential of this cytokine to be exploited as a therapeutic target in atherosclerosis and other inflammatory diseases (Gan et al., 2013).

Although the role of resistin in the vasculature has been an area of intense study, the downstream signalling pathways that mediate resistin actions in VSMC during obesity remain to be uncovered. Accumulating studies have indicated that increased levels of adipokines are important players in the molecular processes operating in the pathogenesis of obesity-mediated cardiovascular disease. Recent studies (Scott et al., 2016) where the effects of diet-induced obesity and resistin on regulation of SGK1 in the vasculature *in vivo* and in VSMC *in vitro* were investigated presented evidence that the elevated plasma resistin that arises in individuals with obesity may determine activation of serum and glucocorticoid-inducible kinase 1 (SGK1). SGK1 is a serine/threonine protein kinase that participates in a wide variety of physiological functions comprising regulation of metabolism, cell proliferation, apoptosis, inflammation, abnormal VSMC proliferation, leading to development of vascular dysfunction. Therefore, it was shown that high resistin-induced stimulation of VSMC proliferation was SGK1-dependent. In addition, these studies suggested that in human SMCs *in vitro*, resistin regulated SGK1 via a phosphatidylinositol-3-kinase (PI3K) and TLR4-dependent mechanism. These findings pointed to inhibition of the resistin/SGK1 pathway as a novel therapeutic target in treating obesity-associated vascular diseases such as atherosclerosis.

Recent studies have explored further high resistin level mechanisms involved in SMC dysfunction and have revealed that systemic resistin correlated to circulating oxidative stress in patients with severe atherosclerosis, induced VSMC dysfunction *in vitro*, and exaggerated intimal hyperplasia *in vivo*. They found that resistin promoted dose- and time-dependent ROS generation and acute-phase inflammatory cytokine production in VSMCs, and that protein kinase ε (PKCε) mediated resistin-induced cellular and molecular effects. The authors concluded: (1) chronic and acute effects of resistin contributed to the novel mechanistic basis for resistin-associated cardiovascular effects; (2) resistin-associated VSMC dysfunction and intimal hyperplasia are related to PKCε-dependent NADPH Oxydase (Nox) activation; and (3) ROS generation targeting the PKCε-Nox pathway may represent a novel strategy in managing resistin-associated atherosclerotic complications (Zuniga et al., 2017).

## 4.3. Effect of High Subendothelial Resistin in Vascular Inflammation

### *4.3.1. Role of Subendothelial Resistin on Vascular Cells, in Monocyte Transmigration and in EC/SMC Cross-Talk*

In the inflamed vascular wall, endothelial and SMCs respond to local cytokine and chemokine signalling, contributing to an accelerated inflammatory process (Hastings et al., 2009). Monocyte transmigration is a key event in initiating atherosclerotic plaque formation; the cytokine and chemokine milieu contribute to this process. Knowing that resistin is a determinant factor in atheroma progression, we hypothesised that this cytokine, when present in the vascular wall, might activate EC and SMC and their interaction and could enhance monocyte infiltration and retention by specific mechanisms.

Considering that resistin was detected at increased levels in the subendothelial space in the atherosclerotic plaque in mice and humans (Jung et al., 2006) and contributed to enhanced macrophage infiltration in the atherosclerotic artery wall (Cho et al., 2011), we investigated the role of subendothelial resistin (sR) in vascular cells activation and cross-talk leading to enhanced monocyte transmigration and we searched for the mechanisms involved (Pirvulescu et al., 2014).

Our recent data show that sR had proinflammatory effects on EC by activating TLR4 and Gi proteins and by inducing TLR4 and MyD88 association in EC and consequently increasing expression of chemokines fractalkine and MCP-1 in EC (Pirvulescu et al., 2014). Moreover, resistin activated TLR4 and Gi proteins in SMC and induced increased expression of fractalkine, MCP-1 and fractalkine receptor (CX3CR1) in SMC (Gan et al., 2013). By investigating the mechanisms operating in resistin in SMC we found and increased oxidative stress (the ROS levels), particularly O2-, activation of NADPH oxidase, increased p22 and NOX1 levels and p38 MAPK, NF-kB, AP-1 and STAT3 activation (Gan et al., 2013; Gan et al., 2013).

Since resistin has proinflammatory effects in vascular cells, we performed studies to evaluate whether resistin interferes also in vascular cells cross-talk. Our new data showed that the communication between EC and SMC enhanced monocyte transmigration, and the chemokines fractalkine and MCP-1 and their receptors (CX3CR1 and CCR2) were functional in monocyte transmigration induced by sR in the EC/SMC system (Pirvulescu et al., 2014). In addition, we have shown that sR produced fractalkine and MCP-1 in the EC/SMC co-culture system by increasing TLR4 and Gi protein activation; TLR4 and Gi were instrumental in monocyte transmigration (Pirvulescu et al., 2014). Moreover, fractalkine was produced by resistin in the EC/SMC system and by acting on CX3CR1 on EC/SMC, contributes to MCP-1 secretion in the system and to enhanced monocyte transmigration; therefore it may be considered an effector protein of resistin. Our data suggest that resistin produced by infiltrated immune cells contribute to the inflammatory process by increasing production of chemokines, leading to progressive recruitment of other monocytes, thus aggravating inflammation in the vessel wall. Our study indicated that resistin interfered in cytokine signalling and could significantly modulate the inflammatory microenvironment within the vascular wall, pointing out new possible targets for therapy to reduce resistin-dependent enhanced macrophage infiltration in the atherosclerotic arterial wall.

## 5. Resistin's Role and Mechanisms in Progression of Atherosclerosis towards Clinical Events

### 5.1. Resistin's Role in Vascular Cells / Immune Cells Cross-Talk

Although a variety of immune cells have been described to be involved in atheroma progression—monocytes, macrophages, T lymphocytes, neutrophils, mastocytes, platelets—monocytes/macrophages have leading roles in modulating inflammation in atherosclerosis. The majority of macrophages derive from recruited monocytes, but it has also been shown that macrophages can proliferate in atherosclerotic plaque (Swirski et al., 2007). The factors that induce monocyte proliferation are not yet elucidated, but a study on Apoe-/- and Ldlr-/- mice suggested that local proliferation is the major source of macrophages in advanced atherosclerotic lesions (Robbins et al., 2013). Also, it has been shown that hypercholesterolemia and defective cholesterol efflux capacity promote excessive hematopoietic stem cells and progenitor cells proliferation in bone marrow, myeloid progenitor cells and monocyte proliferation in spleen. All these processes stimulate monocytosis and the transport of inflammatory monocytes in atherosclerotic lesions (Murphy et al., 2011; Robbins et al., 2012; Westerterp et al., 2012). In this context, macrophage proliferation can be considered a final step in this process of cellular expansion.

As previously stated, SMCs in inflammatory conditions differentiate and migrate from media to intima, where they proliferate and modulate other processes, leading to atheroma progression (Manduteanu and Simionescu, 2012). Thus, SMC and macrophages interact in intima, but have different roles in amplifying the proinflammatory state and in accelerating plaque progression. In atherosclerotic plaque, SMC and infiltrated macrophages interact via direct contact, implying surface molecules like ICAM-1, VCAM-1 and fractalkine and their corresponding receptors, or indirectly, by soluble factors.

Previous studies (Ferri et al., 2012; Ostriker et al., 2014) revealed that the communication between macrophages and SMC modulated inflammation (the cytokinic profile) in macrophages; resistin had specific roles in the process and may influence atheroma evolution. In our recent studies (Tucureanu et al., 2016) we evaluated how macrophage-SMC communication modulated macrophage inflammation (the cytokinic profile), whether fractalkine and resistin may have a role in the process and how macrophage-SMC communication may influence atheroma evolution (Tucureanu et al., 2016). By examining the cytokine profile of 57 cytokines using a Proteome Profiler Array in macrophages subsequent to their interaction with SMC, we found that 81% of the chemokines and 50% of the tested interleukins were up-regulated and that resistin increased predominantly the expression of 12 interleukins (the highest increase: IL-6, IL-27, IL-23 and IL-5) and of GM-CSF, G-CSF, C5/C5a, CD40L, Serpin E1, ICAM-1, MIF and IL-1ra. The highest increase was demonstrated for IL-6 (15-fold), IL-5 (13-fold), IL-27 (12-fold) and IL-23 (10-fold). In addition, in the same experimental conditions, resistin increased the protein expression of nine chemokines; the highest increase was found for MIP-1β (14-fold), CCL1 chemokine (10-fold) and GROα (10-fold). In addition, the expression of Serpin E1 (PAI-inhibitor) was significantly increased (6-fold). Additionally, resistin down-regulated the expression of 10 chemokines, including fractalkine and CXCL16 and of three interleukins, including the inductor of M2 phenotype IL-4 in macrophage-SMC co-culture, with consequences in monocyte chemotaxis. Moreover, macrophage-SMC secretome in presence of resistin enhances monocyte chemotaxis (Pirvulescu et al., 2016). Using Functional Annotation Bioinformatics Microarray Analysis (DAVID), our data showed that nine of the mediators induced by resistin in macrophages interacted with SMC are targets for the canonical signalling pathway NF-kB; 16 of the mediators are targets for AP-1 transcription factor, indicating an extensive proinflammatory state generated by macrophages in the vessel wall.

IL-6 (the highest increase in macrophage-SMC interaction induced by resistin) is an upstream inflammatory cytokine with a major role in transmitting the downstream inflammatory response responsible for atherosclerosis and in the development of CAD (Hartman and Frishman, 2014). It could be presumed that resistin may exacerbate the inflammatory process associated with atherosclerosis, by participating in macrophage activation and by inducing IL-6. Resistin increased in macrophage-SMC cross-talk mainly proinflammatory and pro-atherogenic mediators, but also increased the expression of IL-17E or IL-25, which were shown to have a protective role in atherosclerosis mediated by innate responses (Hirase et al., 2013; Mantani et al., 2015). These data suggest the immunomodulatory role of resistin in macrophages that interacted with SMC. Resistin also increased significantly the expression of Serpin E1, suggesting the pro-thrombotic effect of resistin in macrophages, subsequent to their interaction with SMC.

Our data showed also that CCL1 and GROα were overexpressed by resistin in the macrophage-SMC system by 10-fold compared to macrophage alone; the extensive expression of these mediators induced by resistin is conditioned by macrophage-SMC cross-talk. Previous studies have shown that CCL1 and GROα co-regulate the expression of RANTES (Feuser et al., 2012), mediating leucocyte retention and transendothelial diapedesis (Li and Ley, 2015). Moreover, GROα together with its receptor, CXCR2, contributes to monocyte and neutrophil migration in the subendothelial space (Zhou et al., 2011).

Our data showed that resistin down-regulated the expression of ten chemokines in macrophages interacted with SMC, including fractalkine and CXCL16 and of three interleukins, including the inductor of M2 phenotype, IL-4. Our previous studies showed that subendothelial resistin increased production of fractalkine in a co-culture of endothelial cells-SMC compared to endothelial cells cultured alone (Pirvulescu et al., 2014) and used fractalkine in monocyte transmigration. These new data showed that in the presence of macrophages interacted with SMC, resistin decreased fractalkine expression in macrophages.

All this data showed that resistin has a major role in amplifying the inflammatory profile in the vessel wall, by inducing the expression of cytokines and chemokines in macrophages interacted with SMC. Thus we investigated if these mediators have a functional role in monocyte migration. Our data showed that macrophage-SMC secretome in the presence of resistin enhanced human monocyte chemotaxis, demonstrating the functional role of chemokines produced as a consequence of resistin involvement in the interaction between macrophages and SMC (Tucureanu et al., 2016) (Figure 3).

These new data indicate that resistin may have different effects in different stages of lesion evolution and future experiments will help clarify this matter. Our new data also present evidence of the modulatory roles of fractalkine and resistin on macrophage phenotype in interaction with SMC. Our data may explain the role of SMC interaction in macrophage phenotype modulation as well as the immunomodulatory roles of fractalkine and resistin in the process and may provide new targets for anti-inflammatory therapies in atherosclerosis.

Concerning macrophage-SMC cross-talk, recent data (Zuniga et al., 2017), associate resistin with atherosclerotic plaque vulnerability and showed that resistin determined a change from proliferation to apoptosis in SMC-macrophage co-culture systems, with a cellular composition similar to that found in vulnerable regions of plaques. Resistin effects were found to be mediated by protein kinase C epsilon, indicating that protein kinase C epsilon may represent a therapeutic strategy in resistin-associated atherosclerotic complications.

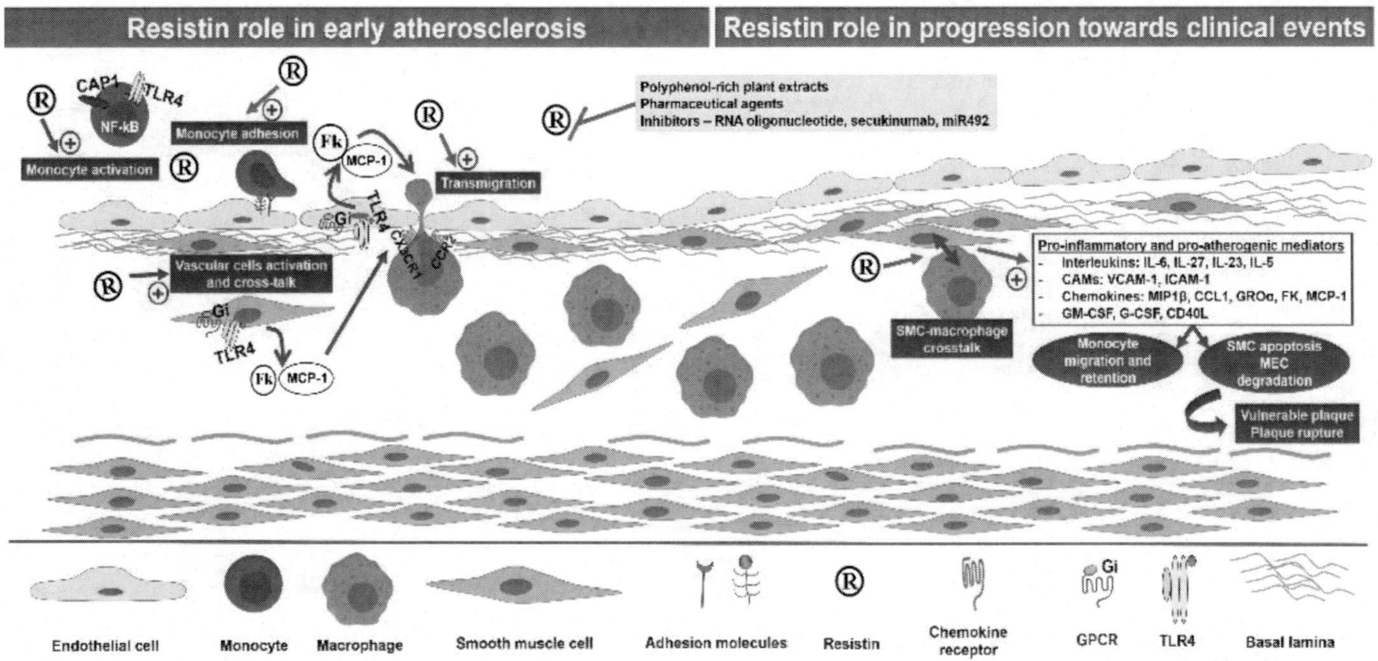


Figure 3. Schematic illustration of mechanism by which resistin enhances inflammatory conditions in the vessel wall and promotes atherosclerotic plaque progression. Resistin may be involved in early stages of atherosclerosis by activating monocytes, endothelial cells and SMC and by increasing monocyte adhesion and transmigration. Within the vessel wall, resistin interferes in cell cross-talk and modulates the expression of proinflammatory and proatherogenic mediators, which may lead to acceleration of plaque progression and rupture. Resistin proinflammatory effects may be reduced by polyphenol-rich plant extracts and pharmaceutical agents and inhibitors such as RNA oligonucleotide, secukinumab and miR-492.

## 6. Resistin Receptor

Resistin mechanisms of action and downstream effects could be properly explained if the receptor for resistin were better characterized. Previous studies proposed decorin (for mouse resistin) and TLR4 and CAP1 (for human resistin) as candidates for resistin receptor (Daquinag et al., 2011; Lee et al., 2014; Tarkowski et al. 2010).

In mouse adipose progenitor cells, resistin interacts with an isoform of decorin that lacks the glycation site; binding of resistin to this peptide promotes cell proliferation and migration but abolishes lipid accumulation after adipogenesis induction, a process mediated by resistin, demonstrating that decorin serves as a functional receptor for resistin (Daquinag et al., 2011).

The first evidence that human resistin binds to TLR4, the innate receptor for lipopolysaccharide, was documented in 2010, when it was shown that resistin competes with LPS for interacting with TLR4 (Tarkowski et al., 2010). Since then, multiple studies, including ours, have investigated the role of TLR4 signalling in resistin effects in various cell types and pathologies. The primary role of TLR4 in innate immunity is to protect the host cells from invading pathogens by generating an intracellular signalling process that activates NF-kB transcription factor, triggering a burst of proinflammatory cytokines and adhesion molecules. Besides its role in innate response, TLR4 has been shown to be activated by endogenous ligands, such as hyaluronan, fibronectin, heat shock proteins, fatty acids and oxidized LDL (Shi et al., 2006; Walton et al., 2003). Previous studies demonstrated the direct binding of resistin to TLR4 in the rat hypothalamus, triggering activation of JNK and p38 MAPK signalling pathways and insulin resistance which revealed strong evidence for the direct role of TLR4 in resistin-induced effects (Benomar et al., 2013).

A recent study pointed out the involvement of a resistin-TLR4 pathway in inflammation, showing that silencing TLR4 abrogated the inflammatory effects of resistin in alveolar macrophages and demonstrating that resistin promotes production of proinflammatory cytokines via a TLR4/NF-kB-mediated pathway (Li et al., 2018). Recent studies reported that increased levels of circulating resistin were correlated with the severity of sepsis, proposing resistin as a diagnostic marker of sepsis (Macdonald et al., 2014). Moreover, transgenic mice that express human resistin showed resistance to LPS-induced inflammation and sepsis, demonstrating the role of resistin in promoting anti-inflammatory pathways such as STAT3 and TBK signalling, with important clinical implications in sepsis (Jang et al., 2017).

Our previous work has shown that resistin modulates the inflammatory process in endothelial cells and SMC by a mechanism involving TLR4 signalling (Gan et al., 2013; Pirvulescu et al., 2012). We also showed that subendothelial resistin activates TLR4 signalling, inducing the secretion of proinflammatory molecules involved in monocyte transmigration (Pirvulescu et al., 2014). Moreover, our results indicate that in vascular cells, resistin may integrate TLR4 and Gi protein signaling, but further experiments are needed to clarify this assumption (Gan et al., 2013; Pirvulescu et al., 2014). In addition, in a study where neutralizing antibodies and siRNA were used to block TLR4, it was shown that this receptor is critical for the expression of CCL19 chemokine in human aortic endothelial cells (Munjas et al., 2017).

There are few studies on the potential role of CAP1 acting as resistin receptor. This concept is based on a study that confirmed the direct binding of resistin to CAP1 in human monocytes, modulating the up-regulation of cAMP concentration, protein kinase A activity and NF-kB activation induced by resistin (Lee et al., 2014). A recent study certified the role of CAP1 in mediating resistin actions in rheumatoid arthritis and CAD (Munjas et al., 2017; Sato et al., 2017).

## 7. Molecules and Mechanisms Involved in Resistin-Dependent Inflammation and Their Potential as Targets for Innovative Nanobiotherapies

Given the proinflammatory effects of serum resistin, it is crucial to find therapies that lower resistin levels and downsize its actions in CVD. Our research and similar studies may help to find new potential targets to reduce resistin-dependent inflammation in vascular cells and in monocytes/ macrophages and to develop novel therapies. Since the discovery of resistin and its implication in various pathologies, anti-resistin therapies were developed using antioxidants, anti-inflammatory agents, antisense oligonucleotides, antibodies and specific inhibitors. But identifying the molecular pathways whereby resistin performs its inflammatory effects will contribute to designing new therapeutic strategies for pharmacological interventions.

## 7.1. Natural Antioxidant Agents Curcumin and Morus Alba Extract Reduce Resistin-Induced Activation of Endothelial Cells

Since our previous data indicated that in endothelial cells, resistin had proinflammatory actions by increasing oxidative stress, we performed studies to investigate whether two natural antioxidants, curcumin and a *Morus alba* extract, counteract the effects of resistin on endothelial cell inflammation.

Curcumin is a yellow pigment from the rhizome of *Curcuma longa* generally used as a spice and food-colouring agent. Curcumin was shown to have anti-inflammatory, antioxidant and cardiovascular protective effects (Aggarwal and Harikumar, 2009; Goel, Kunnumakkara and Aggarwal, 2008; Weisberg, Leibel and Tortoriello, 2008). There are no data on curcumin effects in endothelial cell dysfunction induced by resistin. The extract from *Morus alba* leaves has a unique nutritional profile containing proteins, phenolics, flavonoids and anthocyanins that recommend it as a natural functional tonic (Butt et al., 2008). Mulberry extracts or components, especially flavonoids, have antioxidant (Naowaboot et al., 2009) and anti-inflammatory effects (Enkhmaa et al., 2005; Katsube et al., 2006), providing cardiovascular protection (Butt et al., 2008) and attenuating atherosclerotic lesion development (Enkhmaa et al., 2005). There is no evidence on the modulation of resistin action by *Morus alba* (Pirvulescu et al., 2011).

Since there was no evidence on the modulation of resistin action by *Morus alba*, we evaluated the compound's effects on resistin-induced fractalkine and P-selectin expression and function in monocyte adhesion in human endothelial cells and on resistin-induced increase in intracellular ROS production and NADPH oxidase activation.

Our results showed that *Morus alba* and curcumin reduced resistin-induced P-selectin and fractalkine expression and reduced monocyte adhesion to HEC (Pirvulescu et al., 2011).

To further understand *Morus alba* and curcumin mechanisms of action and considering that we have previously shown that increased ROS production in resistin-stimulated HEC is attributable to NADPH oxidase activity (Manduteanu et al., 2010), we evaluated the potential effects of the compounds on enzyme activity. Our data indicated that *Morus alba* blocks and curcumin attenuates NADPH oxidase activity induced by resistin in HEC.

In conclusion, our results showed that *Morus alba* leaf extract and curcumin reduced the proinflammatory effects of resistin in HEC by inhibiting expression of fractalkine and P-selectin, molecules involved in early and advanced stages of atherosclerosis and in reducing monocyte adhesion. Moreover, our findings indicated that the antioxidant properties of *Morus alba* and curcumin, namely reduction of the increased level of intracellular ROS induced by resistin and NADPH oxidase activity, may be responsible for their beneficial effects in reducing the proinflammatory action of resistin in HEC. In addition, our findings indicated the potential use of *Morus alba* leaf extract and curcumin as therapeutics in counteracting the adverse effects of resistin in human endothelial cells. Our studies highlighted the effects of natural antioxidants such as a *Morus alba* extract or curcumin on the proinflammatory effects of resistin on human endothelial cells. We have highlighted these antioxidants' potential in developing new phytotherapies to correct resistin-induced endothelial dysfunction and to help prevent cardiovascular disease (Pirvulescu et al., 2011).

These data corroborate results in the literature that show the effect of *Morus alba* leaf extract in reduction of serum resistin and glucose levels in rats with streptozotocin-induced type 2 diabetes (Salemi et al., 2016).

## 7.2. Pharmacological Interventions to Reduce Resistin Plasma Levels

Emerging evidence suggests the role of plasma resistin in the inflammatory state. Thus, extensive research is centred on lowering resistin levels and diminishing its actions in CVD.

In a meta-analysis conducted in 2016 regarding intervention trials, it was stated that consumption of plant-based diets is associated with a reduction of CRP, IL-6 and soluble ICAM1 in obese patients, but no substantial effects were seen for TNFα, resistin, adiponectin and leptin (Eichelmann et al., 2016). However, recent studies used phytotherapies in obesity and type 2 diabetes animal models to reduce plasma resistin and to improve the inflammatory profile. In a mouse fed with a high-fat diet supplemented with *Moringa oleifera* (a plant rich in isothiocyanate, with antidiabetic effects), reduction was shown in resistin levels, together with leptin, cholesterol, IL-1β and TNFα plasma levels, suggesting the effect of the plant extract for prevention and treatment of obesity and type 2 diabetes (Waterman et al., 2015). Recent high-fat-diet animal model studies showed that long-term dietary supplementation with yerba mate (*Ilex paraguariensis*) and sea buckthorn leaf extract increased thermogenic gene mRNA expression and energy expenditure, respectively, both correlated with body weight reduction, and decreased plasma levels of free fatty acids, triglycerides, total cholesterol, leptin and resistin. This suggests that these plants are beneficial in improving diet-induced adiposity, insulin resistance and dyslipidaemia (Choi et al., 2017; Kwon et al., 2017). *Vitis vinifera*, rich in polyphenols, was also shown in a high-fat-diet mouse model to have beneficial effects, such as preventing weight gain and dyslipidaemia; insulin resistance; and lowered plasma levels of leptin, adiponectin and resistin. These effects may be related to a positive modulation of insulin signalling proteins and GLUT4 expression and a reduction of proinflammatory markers, such as TNFα and IL-6 (da Costa et al., 2017). Obese diabetic mice fed with red raspberry fruit (rich in polyphenols and fibres) showed protection against diabetes-induced oxidative stress and reduction of total cholesterol, LDL and resistin plasma levels (Noratto, Chew and Atienza, 2017). Extensive studies have utilized polyphenols to reduce plasma levels of inflammatory markers *in vivo*; *in vitro* studies utilizing 3T3-L1 adipocytes showed that LPS enhanced production of TLR4, NF-kB, IL-6, MCP-1, PAI-1 and resistin. But plant polyphenols reversed these effects and increased SOD and catalase antioxidant expression (Le Sage, Meilhac and Gonthier, 2017). Furthermore, catechin and quercetin attenuated adipose inflammation triggered by high fructose in 3T3-L1 adipocytes by reducing expression of resistin and MCP-1 (Vazquez Prieto et al., 2015). Moreover, it was shown that resveratrol, a nutraceutical agent that attracts considerable attention due to its pharmacological potential in a multitude of diseases, reversed diabetes-induced resistin levels and improved hepatic dysfunction induced by diabetes (Sadi et al., 2015).

To reduce plasma resistin and its effects, several research groups employed pharmaceuticals already used in clinics, such as antihypertensive, antidiabetic and anti-inflammatory drugs. A study including 84 obese hypertensive patients treated for six weeks with different antihypertensive drugs showed that plasma resistin was lowered by bisoprolol (β-blocker), amlodipine (calcium antagonist) and indapamide (diuretic drug), but not by angiotensin II receptor type 1 (candesartan). Other adipokines, like apelin and visfatin, showed varied plasma levels after treatment with the antihypertensive drugs; this premise requires further research (Skoczylas, Piecha and Więcek, 2016). A recent study showed that glimepiride, an antidiabetic drug, affected resistin and adiponectin levels in diabetic patients, while metformin influenced plasma leptin (Farooq et al., 2017). Hypercholesterolemic patients with or without associated diabetes were treated with simvastatin and ezetimibe, cholesterol-lowering drugs. In nondiabetic patients, plasma resistin and resistin mRNA expression in peripheral blood mononuclear cells were reduced by add-on ezetimibe therapy, but not in diabetic subjects (Guimarães et al., 2017). *In vitro* studies showed that aspirin, a well-known anti-inflammatory agent, reduced endothelial dysfunction induced by resistin, via modulation of ROS generation and AMPK signalling (Ou et al., 2012).

Novel therapeutic agents such as oligonucleotides and inhibitory antibodies have been used *in vitro* and *in vivo* in various experimental models to reduce resistin effects. It was investigated whether a synthetic RNA oligonucleotide designed to target mouse resistin gene has an effect in a hyperlipidaemic mouse model; possible molecular mechanisms were explored. The study showed that this antiresistin oligonucleotide improved hyperlipidaemia, reduced adipose mass and attenuated overexpression of resistin, TNFα and IL-6 (Tan et al., 2015). Another potential therapeutic agent is secukinumab, a neutralizing monoclonal antibody for IL-17, which has shown potential efficacy in psoriatic arthritis treatment. Since studies showed that adipocytokines involved in chronic inflammation could act by interfering with IL-17 signalling, secukinumab was evaluated for its effects on resistin, chemerin and adiponectin levels. The data showed that only in a male group of 28 psoriatic arthritis patients treated with secukinumab, resistin and chemerin decreased in plasma; but these data are preliminary and lack statistical power due to the limited number of subjects, so further investigation is needed to establish the effects of this antibody on the adipokine profile (Fassio et al., 2017). Last but not least, it was established that a miRNA molecule is responsible for the up-regulation of resistin and endothelial activation induced by high glucose concentration, *in vitro* and *in vivo*. Hence, after high-glucose treatment of apoE-/- mice or HUVEC cells, miR-492 expression significantly decreased, with a notable up-regulation of resistin expression. Using a luciferase reporter assay, it was established that resistin is a direct target of miR-492; thus, up-regulation of this molecule showed a reduction of endothelial dysfunction via directly down-regulating resistin expression (Ying et al., 2014).

While there is an abundance of evidence that plant extracts, pharmaceuticals and various inhibitors could be promising therapeutic agents, clinical trials must confirm their potential. Along with deciphering the molecular pathways whereby resistin interacts in different pathologies, new targets for pharmacological therapy will be identified, thereby preventing inflammatory effects on the cardiovascular system and providing novel strategies in the treatment of vascular diseases.

## CONCLUSION

A growing body of evidence suggests that resistin may have an important role in CVD by interfering in systemic and local inflammation in the vessel wall. Accumulating data suggest that resistin may be involved in disease initiation and progression and in atherosclerotic plaque rupture. Recent data reported that resistin may be involved in the initiation of atherogenesis by activating vascular endothelial cells and SMC and by interfering in their cross-talk, leading to increased monocyte adhesion and infiltration and enhanced vessel wall inflammation (Figure 3). Moreover, resistin, when present in the diseased vessel wall, may orchestrate a cascade of pro-atherosclerotic events, contributing to atherosclerosis progression towards plaque rupture. Thus, resistin may augment monocyte and macrophage infiltration, may interfere in cell cross-talk and may modulate expression of inflammatory cytokines, leading to acceleration of plaque growth and destabilization by mechanisms now being uncovered (Figure 3). Additional work is necessary to identify the specific receptor for human resistin, which would help to uncover signaling pathways and aid in revealing potential targets for new and innovative therapies.

## REFERENCES

Aggarwal, B. B. and Harikumar, K. B. (2009). "Potential Therapeutic Effects of Curcumin, the Anti-Inflammatory Agent, against Neurodegenerative, Cardiovascular, Pulmonary, Metabolic, Auto-immune and Neoplastic Diseases." *The International Journal of Biochemistry & Cell Biology* 41 (1): 40–59. https://doi.org/10.1016/j.biocel.2008.06.010.

Ahn, S. Y., Cho, C. H., Park, K. G., Lee, H. J., Lee, S., Park, S. K., Lee, I. K. and Koh, G. Y. (2004). "Tumor Necrosis Factor-Alpha Induces Fractalkine Expression Preferentially in Arterial Endothelial Cells and Mithramycin A Suppresses TNF-Alpha-Induced Fractalkine Expression." *The American Journal of Pathology* 164 (5): 1663–72.

Al-Aoukaty, A, Rolstad, B., Giaid, A. and Maghazachi, A. A. (1998). "MIP-3alpha, MIP-3beta and Fractalkine Induce the Locomotion and the Mobilization of Intracellular Calcium, and Activate the Heterotrimeric G Proteins in Human Natural Killer Cells." *Immunology* 95 (4): 618–24.

Alexander, M. R. and Owens, G. K. (2012). "Epigenetic Control of Smooth Muscle Cell Differentiation and Phenotypic Switching in Vascular Development and Disease." *Annual Review of Physiology* 74 (1): 13–40. https://doi.org/10.1146/annurev-physiol-012110-142315.

Altannavch, T. S., Roubalová, K., Kucera, P. and Andel, M. (2004). "Effect of High Glucose Concentrations on Expression of ELAM-1, VCAM-1 and ICAM-1 in HUVEC with and without Cytokine Activation." *Physiological Research*/Academia Scientiarum Bohemoslovaca 53 (1): 77–82.

Barnes, K. M. and Miner, J. L. (2009). "Role of Resistin in Insulin Sensitivity in Rodents and Humans." *Current Protein & Peptide Science* 10 (1): 96–107.

Benomar, Y., Gertler, A., De Lacy, P., Crépin, D., Hamouda, H. O., Riffault, L. and Taouis, M. (2013). "Central Resistin Overexposure Induces Insulin Resistance Through Toll-Like Receptor 4." *Diabetes* 62 (1): 102–14. https://doi.org/10.2337/db12-0237.

Bjørbaek, C., El-Haschimi, K., Frantz, J. D. and Flier, J. S. (1999). "The Role of SOCS-3 in Leptin Signaling and Leptin Resistance." *The Journal of Biological Chemistry* 274 (42): 30059–65.

Bluyssen, H. A. R., Rastmanesh, M. M., Tilburgs, C., Jie, K., Wesseling, S., Goumans, M. J., Boer, P., Joles, J. A. and Braam, B. (2010). "IFN-Dependent SOCS3 Expression Inhibits IL-6-Induced STAT3 Phosphorylation and Differentially Affects IL-6 Mediated Transcriptional Responses in Endothelial Cells." *AJP: Cell Physiology* 299 (2): C354–62. https://doi.org/10.1152/ajpcell.00513.2009.

Burnett, M. S., Lee, C. W., Kinnaird, T. D., Stabile, E., Durrani, S., Dullum, M. K. Devaney, J. M., Fishman, C., Stamou, S., Canos, D., Zbinden, S., Clavijo, L. C., Jang, G. J., Andrews, J. A., Zhu, J. and Epstein, S. E. (2005). "The Potential Role of Resistin in Atherogenesis."

*Atherosclerosis* 182 (2): 241–48. https://doi.org/10.1016/j.atherosclerosis.2005.02.014.

Butler, J., Kalogeropoulos, A., Georgiopoulou, V., de Rekeneire, N., Rodondi, N., Smith, A. L., Hoffmann, U., Kanaya, A., Newman, A. B., Kritchevsky, S. B., Vasan, R. S., Wilson, P. W. F. and Harris, T. B. (2009). "Serum Resistin Concentrations and Risk of New Onset Heart Failure in Older Persons: The Health, Aging, and Body Composition (Health ABC) Study." *Arteriosclerosis, Thrombosis, and Vascular Biology* 29 (7): 1144–49. https://doi.org/10.1161/ATVBAHA.109.186783.

Butoi Dragomir, E., Gan, A. M., Manduteanu, I., Stan, D., Calin, M., Pirvulescu, M., Koenen, R. R., Weber, C. and Simionescu, M. (2011). "Cross Talk between Smooth Muscle Cells and Monocytes/Activated Monocytes via CX3CL1/CX3CR1 Axis Augments Expression of pro-Atherogenic Molecules." *Biochimica Et Biophysica Acta*. https://doi.org/10.1016/j.bbamcr.2011.08.009.

Butt, M. S., Nazir, A., Sultan and Schroën, K. (2008). "Morus Alba L. Nature's Functional Tonic." *Trends in Food Science & Technology* 19 (10): 505–12. https://doi.org/10.1016/j.tifs.2008.06.002.

Calabro, P., Samudio, I., Willerson, J. T. and Yeh, E. T. H. (2004). "Resistin Promotes Smooth Muscle Cell Proliferation through Activation of Extracellular Signal-Regulated Kinase 1/2 and Phosphatidylinositol 3-Kinase Pathways." *Circulation* 110 (21): 3335–40. https://doi.org/10.1161/01.CIR.0000147825.97879.E7.

Chapman, G. A., Moores, K., Harrison, D., Campbell, C. A., Stewart, B. R. and Strijbos, P. J. (2000). "Fractalkine Cleavage from Neuronal Membranes Represents an Acute Event in the Inflammatory Response to Excitotoxic Brain Damage." *The Journal of Neuroscience: The Official Journal of the Society for Neuroscience* 20 (15): RC87.

Chen, C., Jiang, J., Lu, J. M., Chai, H., Wang, X., Lin, P. H. and Yao, Q. (2010). "Resistin Decreases Expression of Endothelial Nitric Oxide Synthase through Oxidative Stress in Human Coronary Artery Endothelial Cells." *American Journal of Physiology - Heart and*

*Circulatory Physiology* 299 (1): H193–201. https://doi.org/10.1152/ajpheart.00431.2009.

Cho, Y., Lee, S. E., Lee, H. C., Hur, J., Lee, S., Youn, S. W., Lee, J., Lee, H. J., Lee, T. K., Park, J., Hwang, S. J., Kwon, Y. W., Cho, H. J., Oh, B. H., Park, Y. B. and Kim, H. S. (2011). "Adipokine Resistin Is a Key Player to Modulate Monocytes, Endothelial Cells, and Smooth Muscle Cells, Leading to Progression of Atherosclerosis in Rabbit Carotid Artery." *Journal of the American College of Cardiology* 57 (1): 99–109. https://doi.org/10.1016/j.jacc.2010.07.035.

Choi, M. S., Park, H. J., Kim, S. R., Kim, D. Y. and Jung, U. J. (2017). "Long-Term Dietary Supplementation with Yerba Mate Ameliorates Diet-Induced Obesity and Metabolic Disorders in Mice by Regulating Energy Expenditure and Lipid Metabolism." *Journal of Medicinal Food* 20 (12): 1168–75. https://doi.org/10.1089/jmf.2017.3995.

Codoñer-Franch, P. and Alonso-Iglesias, E. (2015). "Resistin: Insulin Resistance to Malignancy." Clinica Chimica Acta. *International Journal of Clinical Chemistry* 438 (January): 46–54. https://doi.org/10.1016/j.cca.2014.07.043.

da Costa, G. F., Barcellos Santos, I., de Bem, G. F., Cordeiro, V. S. C., da Costa, C. A., de Carvalho, L. C. R. M, Ognibene, D. T., Resende, A. C. and de Moura, R. S. (2017). "The Beneficial Effect of Anthocyanidin-Rich Vitis Vinifera L. Grape Skin Extract on Metabolic Changes Induced by High-Fat Diet in Mice Involves Antiinflammatory and Antioxidant Actions." *Phytotherapy Research* 31 (10): 1621–32. https://doi.org/10.1002/ptr.5898.

Damås, J. K., Boullier, A., Waehre, T., Smith, C., Sandberg, W. J., Green, S., Aukrust, P. and Quehenberger, O. (2005). "Expression of Fractalkine (CX3CL1) and Its Receptor, CX3CR1, Is Elevated in Coronary Artery Disease and Is Reduced during Statin Therapy." *Arteriosclerosis, Thrombosis, and Vascular Biology* 25 (12): 2567–72. https://doi.org/10.1161/01.ATV. 0000190672.36490.7b.

Daquinag, A. C., Zhang, Y., Amaya-Manzanares, F., Simmons, P. J. and Kolonin, M. G. (2011). "An Isoform of Decorin Is a Resistin Receptor on the Surface of Adipose Progenitor Cells." *Cell Stem Cell* 9 (1): 74–86. https://doi.org/10.1016/j.stem.2011.05.017.

Ding, Q., White, S. P., Ling, C. and Zhou, W. (2011). "Resistin and Cardiovascular Disease." *Trends in Cardiovascular Medicine* 21 (1): 20–27. https://doi.org/10.1016/j.tcm.2012.01.004.

Doran, A. C., Meller, N. and McNamara, C. A. (2008). "Role of Smooth Muscle Cells in the Initiation and Early Progression of Atherosclerosis." *Arteriosclerosis, Thrombosis, and Vascular Biology* 28 (5): 812–19. https://doi.org/10.1161/ATVBAHA.107.159327.

Dragomir, E., Tircol, M., Manduteanu, I., Voinea, M. and Simionescu, M. (2006). "Aspirin and PPAR-Alpha Activators Inhibit Monocyte Chemoattractant Protein-1 Expression Induced by High Glucose Concentration in Human Endothelial Cells." *Vascular Pharmacology* 44 (6): 440–49. https://doi.org/10.1016/j.vph.2006.02.006.

Eichelmann, F., Schwingshackl, L., Fedirko, V. and Aleksandrova, K. (2016). "Effect of Plant-Based Diets on Obesity-Related Inflammatory Profiles: A Systematic Review and Meta-Analysis of Intervention Trials." *Obesity Reviews: An Official Journal of the International Association for the Study of Obesity* 17 (11): 1067–79. https://doi.org/10.1111/obr.12439.

Enkhmaa, B., Shiwaku, K., Katsube, T., Kitajima, K., Anuurad, E., Yamasaki, M. and Yamane, Y. (2005). "Mulberry (Morus Alba L.) Leaves and Their Major Flavonol Quercetin 3-(6-Malonylglucoside) Attenuate Atherosclerotic Lesion Development in LDL Receptor-Deficient Mice." *The Journal of Nutrition* 135 (4): 729–34.

Farooq, R., Amin, S., Bhat, M. H., Malik, R., Wani, H. A. and Majid, S. (2017). "Type 2 Diabetes and Metabolic Syndrome - Adipokine Levels and Effect of Drugs." *Gynecological Endocrinology: The Official Journal of the International Society of Gynecological Endocrinology* 33 (1): 75–78. https://doi.org/10.1080/09513590.2016.1207165.

Fassio, A., Gatti, D., Gisondi, P., Girolomoni, G., Viapiana, O., Giollo, A., Zamboni, M., Rossini, M. and Idolazzi, L. (2017). "Effects of Secukinumab on Serum Adipocytokines: Preliminary Data."

*Reumatismo* 69 (3): 105–10. https://doi.org/10.4081/reumatismo.2017.953.

Ferri, N., Tibolla, G., Pirillo, A., Cipollone, F., Mezzetti, A., Pacia, S., Corsini, A. and Catapano, A. L. (2012). "Proprotein Convertase Subtilisin Kexin Type 9 (PCSK9) Secreted by Cultured Smooth Muscle Cells Reduces Macrophages LDLR Levels." *Atherosclerosis* 220 (2): 381–86. https://doi.org/10.1016/j.atherosclerosis.2011.11.026.

Feuser, K., Thon, K. P., Bischoff, S. C. and Lorentz, A. (2012). "Human Intestinal Mast Cells Are a Potent Source of Multiple Chemokines." *Cytokine* 58 (2): 178–85. https://doi.org/10.1016/j.cyto.2012.01.001.

Fong, A. M., Robinson, L. A., Steeber, D. A., Tedder, T. F., Yoshie, O., Imai, T. and Patel, D. D. (1998). "Fractalkine and CX3CR1 Mediate a Novel Mechanism of Leukocyte Capture, Firm Adhesion, and Activation under Physiologic Flow." *The Journal of Experimental Medicine* 188 (8): 1413–19.

Fontana, A., Spadaro, S., Copetti, M., Spoto, B., Salvemini, L., Pizzini, P., Frittitta, L., Mallamaci, F., Pellegrini, F., Trischitta, V. and Menzaghi, C. (2015). "Association between Resistin Levels and All-Cause and Cardiovascular Mortality: A New Study and a Systematic Review and Meta-Analysis." *PloS One* 10 (3): e0120419. https://doi.org/10.1371/journal.pone.0120419.

Frankel, D. S., Vasan, R. S., D'Agostino, R. D., Benjamin, E. J., Levy, D., Wang, T. J. and Meigs, J. B. (2009). "Resistin, Adiponectin, and Risk of Heart Failure. The Framingham Offspring Study." *Journal of the American College of Cardiology* 53 (9): 754–62. https://doi.org/10.1016/j.jacc.2008.07.073.

Fraticelli, P., Sironi, M., Bianchi, G., D'Ambrosio, D., Albanesi, C., Stoppacciaro, A., Chieppa, M., Allavena, P., Ruco, L., Girolomoni, G., Sinigaglia, F., Vecchi, A. and Mantovani, A. (2001). "Fractalkine (CX3CL1) as an Amplification Circuit of Polarized Th1 Responses." *The Journal of Clinical Investigation* 107 (9): 1173–81. https://doi.org/10.1172/JCI11517.

Gan, A. M., Pirvulescu, M. M., Stan, D., Simion, V., Calin, M., Manduteanu, I. and Butoi, E. (2013). "Monocytes and Smooth Muscle Cells Cross-

Talk Activates STAT3 and Induces Resistin and Reactive Oxygen Species Production [Corrected].” *Journal of Cellular Biochemistry* 114 (10): 2273–83. https://doi.org/10.1002/jcb.24571.

Gan, A. M., Dragomir Butoi E., Manea, A., Simion, V., Stan, D., Parvulescu, M. M., Calin, M., Manduteanu, I. and Simionescu, M. (2013). “Inflammatory Effects of Resistin on Human Smooth Muscle Cells: Up-Regulation of Fractalkine and Its Receptor, CX3CR1 Expression by TLR4 and Gi-Protein Pathways.” *Cell and Tissue Research* 351 (1): 161–74. https://doi.org/10.1007/s00441-012-1510-9.

Gasbarrino, K., Mantzoros, C., Gorgui, J., Veinot, J. P., Lai, C. and Daskalopoulou, S. S. (2016). “Circulating Chemerin Is Associated with Carotid Plaque Instability, Whereas Resistin Is Related to Cerebrovascular Symptomatology.” *Arteriosclerosis, Thrombosis, and Vascular* Biology 36 (8): 1670–78. https://doi.org/10.1161/ATVBAHA.115.306741.

Goel, A., Kunnumakkara, A. B. and Aggarwal, B. B. (2008). “Curcumin as ‘Curecumin’: From Kitchen to Clinic.” *Biochemical Pharmacology* 75 (4): 787–809. https://doi.org/10.1016/j.bcp.2007.08.016.

Gualillo, O., González-Juanatey, J. R. and Lago, F. (2007). “The Emerging Role of Adipokines as Mediators of Cardiovascular Function: Physiologic and Clinical Perspectives.” *Trends in Cardiovascular Medicine* 17 (8): 275–83. https://doi.org/10.1016/j.tcm.2007.09.005.

Guimarães, E. S., Cerda, A., Dorea, E. L., Bernik, M. M. S., Gusukuma, M. C., Pinto, G. A., Fajardo, C. M., Hirata, M. H. and Hirata, R. D. C. (2017). “Effects of Short-Term Add-on Ezetimibe to Statin Treatment on Expression of Adipokines and Inflammatory Markers in Diabetic and Dyslipidemic Patients.” *Cardiovascular Therapeutics* 35 (6). https://doi.org/10.1111/1755-5922.12307.

Hartman, J. and Frishman, W. H. (2014). “Inflammation and Atherosclerosis: A Review of the Role of Interleukin-6 in the Development of Atherosclerosis and the Potential for Targeted Drug Therapy.” *Cardiology in Review* 22 (3): 147–51. https://doi.org/10.1097/CRD.0000000000000021.

Hastings, N. E., Feaver, R. E., Lee, M. Y., Wamhoff, B. R. and Blackman, B. R. (2009). "Human IL-8 Regulates Smooth Muscle Cell VCAM-1 Expression in Response to Endothelial Cells Exposed to Atheroprone Flow." *Arteriosclerosis, Thrombosis, and Vascular Biology* 29 (5): 725–31. https://doi.org/10.1161/ATVBAHA.109.184382.

Hirase, T., Hara, H., Miyazaki, Y., Ide, N., Nishimoto-Hazuku, A., Fujimoto, H., Saris, C. J. M., Yoshida, H. and Node, K. (2013). "Interleukin 27 Inhibits Atherosclerosis via Immunoregulation of Macrophages in Mice." *American Journal of Physiology. Heart and Circulatory Physiology* 305 (3): H420–29. https://doi.org/10.1152/ajpheart.00198.2013.

Hogas, S., Bilha, S. C., Branisteanu, D., Hogas, M., Gaipov, A., Kanbay, M. and Covic, A. (2017). "Potential Novel Biomarkers of Cardiovascular Dysfunction and Disease: Cardiotrophin-1, Adipokines and Galectin-3." *Archives of Medical Science: AMS* 13 (4): 897–913. https://doi.org/10.5114/aoms.2016.58664.

Hsu, W. Y., Chao, Y. W., Tsai, Y. L., Lien, C. C., Chang, C. F., Deng, M. C., Ho, L. T., Kwok, C. F. and Juan, C. C. (2011). "Resistin Induces Monocyte-Endothelial Cell Adhesion by Increasing ICAM-1 and VCAM-1 Expression in Endothelial Cells via P38MAPK-Dependent Pathway." *Journal of Cellular Physiology* 226 (8): 2181–88. https://doi.org/10.1002/jcp.22555.

Huang, X., and Yang, Z. (2016). "Resistin's Obesity and Insulin Resistance: The Continuing Disconnect between Rodents and Humans." *Journal of Endocrinological Investigation* 39 (6): 607–15. https://doi.org/10.1007/s40618-015-0408-2.

Imai, T., Hieshima, K., Haskell, C., Baba, M., Nagira, M., Nishimura, M, Kakizaki, M., Takagi, S., Nomiyama, H., Schall, T. J., Yoshie, O. (1997). "Identification and Molecular Characterization of Fractalkine Receptor CX3CR1, Which Mediates Both Leukocyte Migration and Adhesion." *Cell* 91 (4): 521–30.

Jamaluddin, M. S., Weakley, S. M., Yao, Q. and Chen, C. (2012). "Resistin: Functional Roles and Therapeutic Considerations for Cardiovascular

Disease." *British Journal of Pharmacology* 165 (3): 622–32. https://doi.org/ 10.1111/j.1476-5381.2011.01369.x.

Jamaluddin, M. S., Yan, S., Lü, J., Liang, Z., Yao, Q. and Chen, C. (2013). "Resistin Increases Monolayer Permeability of Human Coronary Artery Endothelial Cells." *PloS One* 8 (12): e84576. https://doi.org/10.1371/journal.pone.0084576.

Jang, J. C., Li, J., Gambini, L., Batugedara, H. M., Sati, S., Lazar, M. A., Fan, L., Pellecchia, M. and Nair, M. G. (2017). "Human Resistin Protects against Endotoxic Shock by Blocking LPS-TLR4 Interaction." *Proceedings of the National Academy of Sciences of the United States of America* 114 (48): E10399–408. https://doi.org/10.1073/pnas.1716015114.

Jung, H. S., Park, K. H., Cho, Y. M., Chung, S. S., Cho, H. J., Cho, S. Y., Kim, S. J., Kim, S. Y., Lee, H. K. and Park, K. S. (2006). "Resistin Is Secreted from Macrophages in Atheromas and Promotes Atherosclerosis." *Cardiovascular Research* 69 (1): 76–85. https://doi.org/10.1016/j.cardiores.2005.09.015.

Kashiwagi, A., Asahina, T., Nishio, Y., Ikebuchi, M., Tanaka, Y., Kikkawa, R. and Shigeta, Y. (1996). "Glycation, Oxidative Stress, and Scavenger Activity: Glucose Metabolism and Radical Scavenger Dysfunction in Endothelial Cells." *Diabetes* 45 Suppl 3 (July): S84-86.

Katsube, T., Imawaka, N., Kawano, Y., Yamazaki, Y., Shiwaku, K. and Yamane, Y. (2006). "Antioxidant Flavonol Glycosides in Mulberry (Morus Alba L) Leaves Isolated Based on LDL Antioxidant Activity." *Food Chemistry* 97 (1): 25–31. https://doi.org/10.1016/j.foodchem.2005.03.019.

Kawanami, D., Maemura, K., Takeda, N., Harada, T., Nojiri, T., Imai, Y., Manabe, I., Utsunomiya, K. and Nagai, R. (2004). "Direct Reciprocal Effects of Resistin and Adiponectin on Vascular Endothelial Cells: A New Insight into Adipocytokine-Endothelial Cell Interactions." *Biochemical and Biophysical Research Communications* 314 (2): 415–19.

Kinoshita, K., Furukawa, M., Ebihara, T., Sakurai, A., Noda, A., Kitahata, Y., Utagawa, A. and Tanjoh, K. (2006). "Acceleration of Chemokine

Production from Endothelial Cells in Response to Lipopolysaccharide in Hyperglycemic Condition." *Acta Neurochirurgica. Supplement* 96: 419–21.

Kręçki, R., Krzemińska-Pakuła, M., Peruga, J. Z., Szcześniak, P., Lipiec, P., Wierzbowska-Drabik, K., Orszulak-Michalak, D. and Kasprzak, J. D. (2011). "Elevated Resistin Opposed to Adiponectin or Angiogenin Plasma Levels as a Strong, Independent Predictive Factor for the Occurrence of Major Adverse Cardiac and Cerebrovascular Events in Patients with Stable Multivessel Coronary Artery Disease over 1-Year Follow-Up." *Medical Science Monitor: International Medical Journal of Experimental and Clinical Research* 17 (1): CR26-32.

Kunnari, A. M., Savolainen, E. R., Ukkola, O. H., Antero Kesäniemi, Y. and Jokela, M. A. (2009). "The Expression of Human Resistin in Different Leucocyte Lineages Is Modulated by LPS and TNFalpha." *Regulatory Peptides* 157 (1–3): 57–63. https://doi.org/10.1016/j.regpep.2009.05.002.

Kwon, E. J., Lee, J., Kim, Y. J., Do, A., Choi, J. Y., Cho, S. J., Jung, U. J., Lee, M. K., Park, Y. B. and Choi, M. S. (2017). "Seabuckthorn Leaves Extract and Flavonoid Glycosides Extract from Seabuckthorn Leaves Ameliorates Adiposity, Hepatic Steatosis, Insulin Resistance, and Inflammation in Diet-Induced Obesity." *Nutrients* 9 (6). https://doi.org/10.3390/nu9060569.

Le Sage, F., Meilhac, O. and Gonthier, M. P. (2017). "Anti-Inflammatory and Antioxidant Effects of Polyphenols Extracted from Antirhea Borbonica Medicinal Plant on Adipocytes Exposed to Porphyromonas Gingivalis and Escherichia Coli Lipopolysaccharides." *Pharmacological Research* 119 (May): 303–12. https://doi.org/10.1016/j.phrs.2017.02.020.

Lee, S., Lee, H. C., Kwon, Y. W., Lee, S. E., Cho, Y., Kim, J., Lee, S., Kim, J. Y., Lee, J., Yang, H. M., Mook-Jung, I., Nam, K. Y., Chung, J., Lazar, M. A., Kim, H. S. (2014). "Adenylyl Cyclase-Associated Protein 1 Is a Receptor for Human Resistin and Mediates Inflammatory Actions of Human Monocytes." *Cell Metabolism* 19 (3): 484–97. https://doi.org/10.1016/j.cmet.2014.01.013.

Lee, S.E. and Kim, H. S. (2012). "Human Resistin in Cardiovascular Disease." *Journal of Smooth Muscle Research [Nihon Heikatsukin Gakkai Kikanshi]* 48 (1): 27–35.

Lehrke, M., Reilly, M. P., Millington, S. C., Iqbal, N., Rader, D. J. and Lazar, M. A. (2004). "An Inflammatory Cascade Leading to Hyperresistinemia in Humans." *PLoS Medicine* 1 (2): e45. https://doi.org/10.1371/journal.pmed.0010045.

Lesnik, P., Haskell, C. A. and Charo, I. F. (2003). "Decreased Atherosclerosis in CX3CR1-/- Mice Reveals a Role for Fractalkine in Atherogenesis." *The Journal of Clinical Investigation* 111 (3): 333–40. https://doi.org/10.1172/JCI15555.

Li, B., Fang, J., Zuo, Z., Yin, S., He, T., Yang, M., Deng, J., Shen, L., Ma, X.,Yu, S., Wang, Y., Ren, Z., Cui, H. (2018). "Activation of the Porcine Alveolar Macrophages via Toll-like Receptor 4/NF-KB Mediated Pathway Provides a Mechanism of Resistin Leading to Inflammation." *Cytokine*. https://doi.org/10.1016/j.cyto.2018.04.002.

Li, J. and Ley, K. (2015). "Lymphocyte Migration into Atherosclerotic Plaque." *Arteriosclerosis, Thrombosis, and Vascular Biology* 35 (1): 40–49. https://doi.org/10.1161/ATVBAHA.114.303227.

Liu, K., Liang, C., Liang, Z., Tus, K. and Wakeland, E. K. (2005). "Sle1ab Mediates the Aberrant Activation of STAT3 and Ras-ERK Signaling Pathways in B Lymphocytes." *Journal of Immunology* 174 (3): 1630–37.

Lubos, E., Messow, C. M., Schnabel, R., Rupprecht, H. J., Espinola-Klein, C., Bickel, C., Peetz, D., Post, F., Lackner, K. J., Tiret, L., Münzel, T. and Blankenberg, S. (2007). "Resistin, Acute Coronary Syndrome and Prognosis Results from the AtheroGene Study." *Atherosclerosis* 193 (1): 121–28. https://doi.org/10.1016/j.atherosclerosis.2006.05.039.

Lucas, A. D., Bursill, C., Guzik, T. J., Sadowski, J., Channon, K. M. and Greaves, D. R. (2003). "Smooth Muscle Cells in Human Atherosclerotic Plaques Express the Fractalkine Receptor CX3CR1 and Undergo Chemotaxis to the CX3C Chemokine Fractalkine (CX3CL1)." *Circulation* 108 (20): 2498–2504. https://doi.org/10.1161/01.CIR.0000097119.57756.EF.

Luo, D.; Luo, Y., He, Y., Zhang, H., Zhang, R., Li, X., Dobrucki, W. L., Sinusas, A. J., Sessa, W. C. and Min, W. (2006). "Differential Functions of Tumor Necrosis Factor Receptor 1 and 2 Signaling in Ischemia-Mediated Arteriogenesis and Angiogenesis." *The American Journal of Pathology* 169 (5): 1886–98. https://doi.org/10.2353/ajpath.2006.060603.

Macdonald, S. P. J., Stone, S. F., Neil, C. L., van Eeden, P. E., Fatovich, D. M., Arendts, G. and Brown, S. G. A. (2014). "Sustained Elevation of Resistin, NGAL and IL-8 Are Associated with Severe Sepsis/Septic Shock in the Emergency Department." *PloS One* 9 (10): e110678. https://doi.org/10.1371/journal.pone.0110678.

Manduteanu, I., Voinea, M., Serban, G. and Simionescu, M. (1999). "High Glucose Induces Enhanced Monocyte Adhesion to Valvular Endothelial Cells via a Mechanism Involving ICAM-1, VCAM-1 and CD18." *Endothelium: Journal of Endothelial Cell Research* 6 (4): 315–24.

Manduteanu, I., Dragomir, E., Calin, M., Pirvulescu, M., Gan, A. M., Stan, D. and Simionescu, M. (2009). "Resistin Up-Regulates Fractalkine Expression in Human Endothelial Cells: Lack of Additive Effect with TNF-Alpha." *Biochemical and Biophysical Research Communications* 381 (1): 96–101. https://doi.org/10.1016/j.bbrc.2009.02.015.

Manduteanu, I., Dragomir, E., Voinea, M., Capraru, M. and Simionescu, M. (2007). "Enoxaparin Reduces $H_2O_2$-Induced Activation of Human Endothelial Cells by a Mechanism Involving Cell Adhesion Molecules and Nuclear Transcription Factors." *Pharmacology* 79 (3): 154–62. https://doi.org/10.1159/000098952.

Manduteanu, I., Pirvulescu, M., Gan, A. M., Stan, D., Simion, V., Dragomir, E., Calin, M., Manea, A. and Simionescu, M. (2010). "Similar Effects of Resistin and High Glucose on P-Selectin and Fractalkine Expression and Monocyte Adhesion in Human Endothelial Cells." *Biochemical and Biophysical Research Communications* 391 (3): 1443–48. https://doi.org/10.1016/j.bbrc.2009.12.089.

Manduteanu, I. and Simionescu, M. (2012). "Inflammation in Atherosclerosis: A Cause or a Result of Vascular Disorders?" *Journal of Cellular and Molecular Medicine* 16 (9): 1978–90. https://doi.org/10.1111/j.1582-4934.2012.01552.x.

Mantani, P. T., Dunér, P., Bengtsson, E., Alm, R., Ljungcrantz, I., Söderberg, I., Sundius, L., To, F., Nilsson, J., Björkbacka, H. and Frederikson, G. N. (2015). "IL-25 Inhibits Atherosclerosis Development in Apolipoprotein E Deficient Mice." *PloS One* 10 (1): e0117255. https://doi.org/10.1371/journal.pone.0117255.

McDermott, D. H., Fong, A. M., Yang, Q., Sechler, J. M., Cupples, L. A., Merrell, M. N., Wilson, P. W., D'Agostino, R. B., O'Donnell, C. J., Patel, D. D. and Murphy, P. M. (2003). "Chemokine Receptor Mutant CX3CR1-M280 Has Impaired Adhesive Function and Correlates with Protection from Cardiovascular Disease in Humans." *Journal of Clinical Investigation* 111 (8): 1241–50. https://doi.org/10.1172/JCI200316790.

Menzaghi, C., Marucci, A., Antonucci, A., De Bonis, C., Ortega Moreno, L., Salvemini, L., Copetti, M., Trischitta, V. and Di Paola, R. (2017). "Suggestive Evidence of a Multi-Cytokine Resistin Pathway in Humans and Its Role on Cardiovascular Events in High-Risk Individuals." *Scientific Reports* 7 (March): 44337. https://doi.org/10.1038/srep44337.

Migita, K., Maeda, Y., Miyashita, T., Kimura, H., Nakamura, M., Ishibashi, H. and Eguchi, K. (2006). "The Serum Levels of Resistin in Rheumatoid Arthritis Patients." *Clinical and Experimental Rheumatology* 24 (6): 698–701.

Munjas, J., Sopić, M., Spasojević-Kalimanovska, V., Kalimanovska-Oštrić, D., Anđelković, K. and Jelić-Ivanović, Z. (2017). "Association of Adenylate Cyclase-Associated Protein 1 with Coronary Artery Disease." *European Journal of Clinical Investigation* 47 (9): 659–66. https://doi.org/10.1111/eci.12787.

Murphy, A. J., Akhtari, M., Tolani, S., Pagler, T., Bijl, N., Kuo, C. L., Wang, M., Sanson, M., Abramowicz, S., Welch, C., Bochem, A. E., Kuivenhoven, J. A., Yvan-Charvet, L. and Tall, A. R. (2011). "ApoE Regulates Hematopoietic Stem Cell Proliferation, Monocytosis, and Monocyte Accumulation in Atherosclerotic Lesions in Mice." *The*

*Journal of Clinical Investigation* 121 (10): 4138–49. https://doi.org/10.1172/JCI57559.

Muse, E. D., Feldman, D. I., Blaha, M. J., Dardari, Z., Blumenthal, R. S., Budoff, M. J., Nasir, K., Criqui, M. H., Cushman, M., McClelland, R. L. and Allison, M. A. (2015). "The Association of Resistin with Cardiovascular Disease in the Multi-Ethnic Study of Atherosclerosis." *Atherosclerosis* 239 (1): 101–8. https://doi.org/10.1016/j.atherosclerosis.2014.12.044.

Nagaev, I., Bokarewa, M., Tarkowski, A. and Smith, U. (2006). "Human Resistin Is a Systemic Immune-Derived Proinflammatory Cytokine Targeting Both Leukocytes and Adipocytes." *PloS One* 1: e31. https://doi.org/10.1371/journal.pone.0000031.

Naowaboot, J., Pannangpetch, P., Kukongviriyapan, V., Kongyingyoes, B. and Kukongviriyapan, U. (2009). "Antihyperglycemic, Antioxidant and Antiglycation Activities of Mulberry Leaf Extract in Streptozotocin-Induced Chronic Diabetic Rats." *Plant Foods for Human Nutrition (Dordrecht, Netherlands)* 64 (2): 116–21. https://doi.org/10.1007/s11130-009-0112-5.

Noratto, G. D., Chew, B. P. and Atienza, L. M. (2017). "Red Raspberry (Rubus Idaeus L) Intake Decreases Oxidative Stress in Obese Diabetic (Db/Db) Mice." *Food Chemistry* 227 (July): 305–14. https://doi.org/10.1016/j.foodchem.2017.01.097.

On, Y. K., Park, H. K., Hyon, M. S. and Jeon, E. S. (2007). "Serum Resistin as a Biological Marker for Coronary Artery Disease and Restenosis in Type 2 Diabetic Patients." *Circulation Journal: Official Journal of the Japanese Circulation Society* 71 (6): 868–73.

Ortiz-Munoz, G., Martin-Ventura, J. L., Hernandez-Vargas, P., Mallavia, B., Lopez-Parra, V., Lopez-Franco, O., Munoz-Garcia, B., Fernandez-Vizarra, P., Ortega, L., Egido, L., Egido, J. and Gomez-Guerrero, C. (2009). "Suppressors of Cytokine Signaling Modulate JAK/STAT-Mediated Cell Responses During Atherosclerosis." *Arteriosclerosis, Thrombosis, and Vascular Biology* 29 (4): 525–31. https://doi.org/10.1161/ATVBAHA.108.173781.

Osawa, H., Doi, Y., Makino, H., Ninomiya, T., Yonemoto, K., Kawamura, R., Hata, J., Tanizaki, Y., Iida, M. and Kiyohara, Y. (2009). "Diabetes and Hypertension Markedly Increased the Risk of Ischemic Stroke Associated with High Serum Resistin Concentration in a General Japanese Population: The Hisayama Study." *Cardiovascular Diabetology* 8: 60. https://doi.org/10.1186/1475-2840-8-60.

Ostriker, A., Horita, H. N., Poczobutt, J., Weiser-Evans, M. C. M. and Nemenoff, R. A. (2014). "Vascular Smooth Muscle Cell-Derived Transforming Growth Factor-β Promotes Maturation of Activated, Neointima Lesion-like Macrophages." *Arteriosclerosis, Thrombosis, and Vascular Biology* 34 (4): 877–86. https://doi.org/10.1161/ATVBAHA.114.303214.

O'Sullivan, L. A., Liongue, C., Lewis, R. S., Stephenson, S. E. M. and Ward, A. C. (2007). "Cytokine Receptor Signaling through the Jak-Stat-Socs Pathway in Disease." *Molecular Immunology* 44 (10): 2497–2506. https://doi.org/10.1016/j.molimm.2006.11.025.

Ou, H. C., Lee, W. J., Wu, C. M., Chen, J. F. M. and Sheu, W. H. H. (2012). "Aspirin Prevents Resistin-Induced Endothelial Dysfunction by Modulating AMPK, ROS, and Akt/ENOS Signaling." *Journal of Vascular Surgery* 55 (4): 1104–15. https://doi.org/10.1016/j.jvs.2011.10.011.

Owens, C. D., Kim, J. M., Hevelone, N. D., Hamdan, A., Raffetto, J. D., Creager, M. A. and Conte, M. S. (2010). "Novel Adipokines, High Molecular Weight Adiponectin and Resistin, Are Associated with Outcomes Following Lower Extremity Revascularization with Autogenous Vein." *Journal of Vascular Surgery* 51 (5): 1152–59. https://doi.org/10.1016/j.jvs.2009.12.051.

Park, H. K. and Ahima, R. S. (2013). "Resistin in Rodents and Humans." *Diabetes & Metabolism Journal* 37 (6): 404–14. https://doi.org/10.4093/dmj.2013.37.6.404.

Park, H. K., Kwak, M. K., Kim, H. J. and Ahima, R. S. (2017). "Linking Resistin, Inflammation, and Cardiometabolic Diseases." *The Korean Journal of Internal Medicine* 32 (2): 239–47. https://doi.org/10.3904/kjim.2016.229.

Pine, G. M., Batugedara, H. M. and Nair, M. G. (2018). "Here, There and Everywhere: Resistin-like Molecules in Infection, Inflammation, and Metabolic Disorders." *Cytokine*. https://doi.org/10.1016/j.cyto.2018.05.014.

Pirvulescu, M. M., Gan, A. M., Stan, D., Simion, V., Calin, M., Butoi, E. and Manduteanu, I. (2014). "Subendothelial Resistin Enhances Monocyte Transmigration in a Co-Culture of Human Endothelial and Smooth Muscle Cells by Mechanisms Involving Fractalkine, MCP-1 and Activation of TLR4 and Gi/o Proteins Signaling." *The International Journal of Biochemistry & Cell Biology* 50 (May): 29–37. https://doi.org/ 10.1016/j.biocel.2014.01.022.

Pirvulescu, M. M., Gan, A. M., Stan, D., Simion, V., Calin, M., Butoi, E., Ionescu Tirgoviste, C. and Manduteanu, I. (2011). "Curcumin and a Morus Alba Extract Reduce Pro-Inflammatory Effects of Resistin in Human Endothelial Cells." *Phytotherapy Research: PTR*. https://doi.org/10.1002/ptr.3463.

Pirvulescu, M., Manduteanu, I., Gan, A. M., Stan, D., Simion, V., Butoi, E., Calin, M. and Simionescu, M. (2012). "A Novel Pro-Inflammatory Mechanism of Action of Resistin in Human Endothelial Cells: Up-Regulation of SOCS3 Expression through STAT3 Activation." *Biochemical and Biophysical Research Communications* 422 (2): 321–26. https://doi.org/10.1016/j.bbrc.2012.04.159.

Qiao, X. Z., Yang, Y. M., Xu, Z.R. and Yang, L. A. (2007). "Relationship between Resistin Level in Serum and Acute Coronary Syndrome or Stable Angina Pectoris." *Journal of Zhejiang University. Science*. B 8 (12): 875–80. https://doi.org/10.1631/jzus.2007.B0875.

Reilly, M. P., Lehrke, M., Wolfe, M. L., Rohatgi, A., Lazar, M. A. and Rader, D. J. (2005). "Resistin Is an Inflammatory Marker of Atherosclerosis in Humans." *Circulation* 111 (7): 932–39. https://doi.org/10.1161/01.CIR.0000155620.10387.43.

Robbins, C. S., Chudnovskiy, A., Rauch, P. J., Figueiredo, J. L., Iwamoto, Y., Gorbatov, R., Etzrodt, M, Weber, G. F., Ueno, T., van Rooijen, N., Mulligan-Kehoe, M. J., Libby, P., Nahrendorf, M., Pittet, M. J., Weissleder, R. and Swirski, F. K. (2012). "Extramedullary

Hematopoiesis Generates Ly-6C(High) Monocytes That Infiltrate Atherosclerotic Lesions." *Circulation* 125 (2): 364–74. https://doi.org/10.1161/CIRCULATIONAHA.111.061986.

Robbins, C. S., Hilgendorf, I., Weber, G. F., Theurl, I., Iwamoto, Y., Figueiredo, J. L., Gorbatov, R., Sukhova, G. K., Gerhardt, L. M., Smyth, D., Zavitz, C. C., Shikatani, E. A., Parsons, M., van Rooijen, N., Lin, H. Y., Husain, M., Libby, P., Nahrendorf, M., Weisslender, R. and Swirski, F. K. (2013). "Local Proliferation Dominates Lesional Macrophage Accumulation in Atherosclerosis." *Nature Medicine* 19 (9): 1166–72. https://doi.org/10.1038/nm.3258.

Ruth, J. H., Volin, M. V., Haines, G. K., Woodruff, D. C., Katschke, K. J., Woods, J. M., Park, C. C., Morel, J. C. M. and Koch, A. E. (2001). "Fractalkine, a Novel Chemokine in Rheumatoid Arthritis and in Rat Adjuvant-Induced Arthritis." *Arthritis & Rheumatism* 44 (7): 1568–81. https://doi.org/10.1002/1529-0131(200107)44:7<1568::AID-ART280>3.0.CO;2-1.

Sadi, G., Pektaş, M. B., Koca, H. B., Tosun, M. and Koca, T. (2015). "Resveratrol Improves Hepatic Insulin Signaling and Reduces the Inflammatory Response in Streptozotocin-Induced Diabetes." *Gene* 570 (2): 213–20. https://doi.org/10.1016/j.gene.2015.06.019.

Salemi, Z., Barzin Tond, S., Fallah, S., Shojaii, A. and Seifi, M. (2016). "The Effect of Morus Alba Leaves Extract and Powder on Resistin Levels and Liver Transaminase Enzymes Activities in Diabetes." *Cellular and Molecular Biology* 62 (6): 112–18.

Sato, H., Muraoka, S., Kusunoki, N., Masuoka, S., Yamada, S., Ogasawara, H., Imai, T., Akasaka, Y., Tochigi, N., Takahashi, H., Tsuchiya, K., Kawai, S. and Nanki, T. (2017). "Resistin Upregulates Chemokine Production by Fibroblast-like Synoviocytes from Patients with Rheumatoid Arthritis." *Arthritis Research & Therapy* 19 (1): 263. https://doi.org/10.1186/s13075-017-1472-0.

Schulz, C., Schäfer, A., Stolla, M., Kerstan, S., Lorenz, M.,von Brühl, M. L., Schiemann, M., Bauersachs, J., Gloe, T., Busch, D. H., Gawaz, M. and Massberg, S. (2007). "Chemokine Fractalkine Mediates Leukocyte Recruitment to Inflammatory Endothelial Cells in Flowing Whole

Blood: A Critical Role for P-Selectin Expressed on Activated Platelets." *Circulation* 116 (7): 764–73. https://doi.org/10.1161/CIRCULATIONAHA.107.695189.

Schwartz, D. R. and Lazar, M. A. (2011). "Human Resistin: Found in Translation from Mouse to Man." *Trends in Endocrinology and Metabolism: TEM* 22 (7): 259–65. https://doi.org/10.1016/j.tem.2011.03.005.

Scott, T. A., Babayeva, O., Banerjee, S., Zhong, W. and Francis, S. C. (2016). "SGK1 Activity Is Modulated by Resistin in Vascular Smooth Muscle Cells and in the Aorta in Response to Diet-Induced Obesity." *Obesity* 24 (3): 678–86. https://doi.org/10.1002/oby.21425.

Shi, H., Kokoeva, M. V., Inouye, K., Tzameli, I., Yin, H. and Flier, J. S. (2006). "TLR4 Links Innate Immunity and Fatty Acid–Induced Insulin Resistance." *The Journal of Clinical Investigation* 116 (11): 3015–25. https://doi.org/10.1172/JCI28898.

Singh, S., Chouhan, S., Mohammad, N. and Bhat, M. K. (2017). "Resistin Causes G1 Arrest in Colon Cancer Cells through Upregulation of SOCS3." *FEBS Letters* 591 (10): 1371–82. https://doi.org/10.1002/1873-3468.12655.

Skoczylas, A., Piecha, G. and Więcek, A. (2016). "Effects of Antihypertensive Treatment on Plasma Apelin, Resistin, and Visfatin Concentrations." *Polskie Archiwum Medycyny Wewnetrznej* 126 (4): 243–53. https://doi.org/10.20452/pamw.3390.

Skoda, M., Stangret, A. and Szukiewicz, D. (2018). "Fractalkine and Placental Growth Factor: A Duet of Inflammation and Angiogenesis in Cardiovascular Disorders." *Cytokine & Growth Factor Reviews* 39 (February): 116–23. https://doi.org/10.1016/j.cytogfr.2017.12.001.

Steppan, C. M., Bailey, S. T., Bhat, S., Brown, E. J., Banerjee, R. R., Wright, C. M., Patel, H. R., Ahima, R. S. and Lazar, M. A. (2001). "The Hormone Resistin Links Obesity to Diabetes." *Nature* 409 (6818): 307–12. https://doi.org/10.1038/35053000.

Steppan, C. M., Wang, J., Whiteman, E. L., Birnbaum, M. J. and Lazar, M. A. (2005). "Activation of SOCS-3 by Resistin." *Molecular and Cellular*

*Biology* 25 (4): 1569–75. https://doi.org/10.1128/MCB.25.4.1569-1575.2005.

Stolla, M., Pelisek, J., von Brühl, M. L., Schäfer, A., Barocke, V., Heider, P., Lorenz, M., Tirniceriu, A., Steinhart, A., Bauersachs, J., Bray, P. F., Massberg, S. and Schultz, C. (2012). "Fractalkine Is Expressed in Early and Advanced Atherosclerotic Lesions and Supports Monocyte Recruitment via CX3CR1." *PloS One* 7 (8): e43572. https://doi.org/10.1371/journal.pone.0043572.

Swirski, F. K., Libby, P., Aikawa, E., Alcaide, P., Luscinskas, F. W., Weissleder, R. and Pittet, M. J. (2007). "Ly-6Chi Monocytes Dominate Hypercholesterolemia-Associated Monocytosis and Give Rise to Macrophages in Atheromata." *The Journal of Clinical Investigation* 117 (1): 195–205. https://doi.org/10.1172/JCI29950.

Tan, Y., Jin, X. L., Lao, W., Kim, J., Xiao, L. and Qu, X. (2015). "Antiresistin RNA Oligonucleotide Ameliorates Diet-Induced Nonalcoholic Fatty Liver Disease in Mice through Attenuating Proinflammatory Cytokines." *BioMed Research International* 2015: 414860. https://doi.org/10.1155/2015/414860.

Tarkowski, A., Bjersing, J., Shestakov, A. and Bokarewa, M. I. (2010). "Resistin Competes with Lipopolysaccharide for Binding to Toll-like Receptor 4." *Journal of Cellular and Molecular Medicine* 14 (6B): 1419–31. https://doi.org/10.1111/j.1582-4934.2009.00899.x.

Tucureanu, M. M., Butoi, E., Gan, A. M., Stan, D., Constantinescu, C. A., Calin, M., Simionescu, M. and Manduteanu, I. (2016). "Amendment of the Cytokine Profile in Macrophages Subsequent to Their Interaction with Smooth Muscle Cells: Differential Modulation by Fractalkine and Resistin." *Cytokine* 83 (July): 250–61. https://doi.org/10.1016/j.cyto.2016.04.019.

Umehara, H., Bloom, E. T., Okazaki, T., Nagano, Y., Yoshie, O. and Imai, T. (2004). "Fractalkine in Vascular Biology: From Basic Research to Clinical Disease." *Arteriosclerosis, Thrombosis, and Vascular Biology* 24 (1): 34–40. https://doi.org/10.1161/01.ATV.0000095360.62479.1F.

Vazquez Prieto, M. A., Bettaieb, A., Rodriguez Lanzi, C., Soto, V. C., Perdicaro, D. J., Galmarini, C. R., Haj, F. G., Miatello, R. M. and Oteiza,

P. I. (2015). "Catechin and Quercetin Attenuate Adipose Inflammation in Fructose-Fed Rats and 3T3-L1 Adipocytes." *Molecular Nutrition & Food Research* 59 (4): 622–33. https://doi.org/10.1002/mnfr.201400631.

Verma, S., Li, S. H., Wang, C. H., Fedak, P. W. M., Li, R. K., Weisel, R. D. and Mickle, D. A. G. (2003). "Resistin Promotes Endothelial Cell Activation: Further Evidence of Adipokine-Endothelial Interaction." *Circulation* 108 (6): 736–40. https://doi.org/10.1161/01.CIR.0000084503.91330.49.

Vozarova de Courten, B., Degawa-Yamauchi, M., Considine, R. V., Tataranni, P. A. and Volarova de Courten, B. (2004). "High Serum Resistin Is Associated with an Increase in Adiposity but Not a Worsening of Insulin Resistance in Pima Indians." *Diabetes* 53 (5): 1279–84.

Walton, K. A., Hsieh, X., Gharavi, N., Wang, S., Wang, G., Yeh, M., Cole, A. L. and Berliner, J. A. (2003). "Receptors Involved in the Oxidized 1-Palmitoyl-2-Arachidonoyl-Sn-Glycero-3-Phosphorylcholine-Mediated Synthesis of Interleukin-8. A Role For Toll-Like Receptor 4 and a GlycosylphoSphatidylinositol-Anchored Protein." *Journal of Biological Chemistry* 278 (32): 29661–66. https://doi.org/10.1074/jbc.M300738200.

Wang, H., Chen, D. Y., Cao, J., He, Z. Y., Zhu, B. P. and Long, M. (2009). "High Serum Resistin Level May Be an Indicator of the Severity of Coronary Disease in Acute Coronary Syndrome." *Chinese Medical Sciences Journal [Chung-Kuo I Hsueh K'o Hsueh Tsa Chih]* 24 (3): 161–66.

Wang, J. H., Lee, C. J., Yang, C. F., Chen, Y. C. and Hsu, B. G. (2017). "Serum Resistin as an Independent Marker of Aortic Stiffness in Patients with Coronary Artery Disease." *PloS One* 12 (8): e0183123. https://doi.org/10.1371/journal.pone.0183123.

Waterman, C., Rojas-Silva, P., Boyunegmez Tumer, T., Kuhn, P., Richard, A. J., Wicks, S., Stephens, J. M., Wang, Z., Mynatt, R., Cefalu, W. and Raskin, I. (2015). "Isothiocyanate-Rich Moringa Oleifera Extract Reduces Weight Gain, Insulin Resistance, and Hepatic Gluconeogenesis in Mice." *Molecular Nutrition & Food Research* 59 (6): 1013–24. https://doi.org/10.1002/mnfr.201400679.

Weikert, C., Westphal, S., Berger, K., Dierkes, J., Möhlig, M., Spranger, J., Rimm, E. B., Willich, S. N., Boeing, H. and Pischon, T. (2008). "Plasma Resistin Levels and Risk of Myocardial Infarction and Ischemic Stroke." *The Journal of Clinical Endocrinology and Metabolism* 93 (7): 2647–53. https://doi.org/10.1210/jc.2007-2735.

Weisberg, S. P., Leibel, R. and Tortoriello, D. V. (2008). "Dietary Curcumin Significantly Improves Obesity-Associated Inflammation and Diabetes in Mouse Models of Diabesity." *Endocrinology* 149 (7): 3549–58. https://doi.org/10.1210/en.2008-0262.

Westerterp, M., Gourion-Arsiquaud, S., Murphy, A. J., Shih, A., Cremers, S., Levine, R. L., Tall, A. R. and Yvan-Charvet, L. (2012). "Regulation of Hematopoietic Stem and Progenitor Cell Mobilization by Cholesterol Efflux Pathways." *Cell Stem Cell* 11 (2): 195–206. https://doi.org/10.1016/j.stem.2012.04.024.

Wolk, R., Berger, P., Lennon, R. J., Brilakis, E. S., Johnson, B. D. and Somers, V. K. (2004). "Plasma Leptin and Prognosis in Patients with Established Coronary Atherosclerosis." *Journal of the American College of Cardiology* 44 (9): 1819–24. https://doi.org/10.1016/j.jacc.2004.07.050.

Xia, L., Pan, J., Yao, L. and McEver, R. P. (1998). "A Proteasome Inhibitor, an Antioxidant, or a Salicylate, but Not a Glucocorticoid Blocks Constitutive and Cytokine-Inducible Expression of P-Selectin in Human Endothelial Cells." *Blood* 91 (5): 1625–32.

Xu, W., Yu, L., Zhou, W. and Luo, M. (2006). "Resistin Increases Lipid Accumulation and CD36 Expression in Human Macrophages." *Biochemical and Biophysical Research Communications* 351 (2): 376–82. https://doi.org/10.1016/j.bbrc.2006.10.051.

Yajima, N., Kasama, T., Isozaki, T., Odai, T., Matsunawa, M., Negishi, M., Ide, H., Kameoka, Y., Hirohata, S. and Adachi, M. (2005). "Elevated Levels of Soluble Fractalkine in Active Systemic Lupus Erythematosus: Potential Involvement in Neuropsychiatric Manifestations." *Arthritis and Rheumatism* 52 (6): 1670–75. https://doi.org/10.1002/art.21042.

Yanavitski, M. and Givertz, M. M. (2011). "Novel Biomarkers in Acute Heart Failure." *Current Heart Failure Reports* 8 (3): 206–11. https://doi.org/10.1007/s11897-011-0065-5.

Yang, B., Li, M., Chen, B. and Li, T. D. (2012). "Resistin Involved in Endothelial Dysfunction among Preclinical Tibetan Male Young Adults." *Journal of the Renin-Angiotensin-Aldosterone System: JRAAS* 13 (4): 420–25. https://doi.org/10.1177/1470320312444745.

Yang, W., Li, Y., Wang, J. Y., Han, R. and Wang, L. (2018). "Circulating Levels of Adipose Tissue-Derived Inflammatory Factors in Elderly Diabetes Patients with Carotid Atherosclerosis: A Retrospective Study." *Cardiovascular Diabetology* 17 (1): 75. https://doi.org/10.1186/s12933-018-0723-y.

Ying, C., Sui-Xin, L., Kang-Ling, X., Wen-Liang, Z., Lei, D., Yuan, L., Fan, Z. and Chen, Z. (2014). "MicroRNA-492 Reverses High Glucose-Induced Insulin Resistance in HUVEC Cells through Targeting Resistin." *Molecular and Cellular Biochemistry* 391 (1–2): 117–25. https://doi.org/10.1007/s11010-014-1993-7.

Zhang, M. H., Na, B., Schiller, N. B. and Whooley, M. A. (2011). "Association of Resistin with Heart Failure and Mortality in Patients with Stable Coronary Heart Disease: Data from the Heart and Soul Study." *Journal of Cardiac Failure* 17 (1): 24–30. https://doi.org/10.1016/j.cardfail.2010.08.007.

Zhou, Z., Subramanian, P., Sevilmis, G., Globke, B., Soehnlein, O., Karshovska, E., Megens, R., Heyll, K., Chun, J., Saulnier-Blache, J. S., Reinholz, M., van Zandvoort, M., Weber, C. and Schober, A. (2011). "Lipoprotein-Derived Lysophosphatidic Acid Promotes Atherosclerosis by Releasing CXCL1 from the Endothelium." *Cell Metabolism* 13 (5): 592–600. https://doi.org/10.1016/j.cmet.2011.02.016.

Zuniga, M. C., Raghuraman, G., Hitchner, E., Weyand, C., Robinson, W. and Zhou, W. (2017). "PKC-Epsilon and TLR4 Synergistically Regulate Resistin-Mediated Inflammation in Human Macrophages." *Atherosclerosis* 259 (April): 51–59. https://doi.org/10.1016/j.atherosclerosis.2017.02.021.

In: Resistin
Editor: Gerald Maldonado
ISBN: 978-1-53614-543-4


*Chapter 5*

# RESISTIN IN ACUTE PANCREATITIS

***Samer AL Sawalhi*[*], *MD***
Department of Surgery, Taibah University-Al-Madina,
Saudi Arabia

## ABSTRACT

*Introduction:* This chapter deals with an important issue, based on the known sequential events of obesity preceding severe acute pancreatitis (AP). The task of investigating a potential new predictive marker of fat necrosis and pancreatitis severity will provide a useful clinical tool. This novel adipose-driven protein (resistin), produced by pancreatic islets, may represent a useful additional marker for the severity of AP.

*Methods:* This prospective study was conducted over three consecutive years. We included 112 patients who were diagnosed with acute biliary pancreatitis. We measured body mass index (BMI), waist circumference (WC), and serum resistin levels by enzyme-linked immunosorbent assay (ELISA) and compared this with the control group having similar gender, BMI, and age.

*Results:* A total of 112 patients were diagnosed with acute gallstone-induced pancreatitis. The mean BMI value was 31.4 kg/m2 (obese, class 1). Twenty seven patients (24.1%) had severe AP, while eighty-five

[*] Corresponding Author Email: drsawalhi@yahoo.com.

patients had mild pancreatitis (75.9%). Our results showed that the BMI significantly correlated with pancreatitis severity (P = 0.007). Serum resistin levels were significantly higher in patients with AP when compared with control subjects (P < 0.0001).

*Conclusion:* It was demonstrated that there is a correlation between serum resistin and AP, and therefore it can be used as a diagnostic marker for AP.

## 1. Introduction

Acute pancreatitis (AP) is potentially life-threatening acute inflammatory condition of the pancreas. Although most cases are mild and resolve spontaneously. It is a disease with an overall mortality rate ranging between 4% to 40% in severe cases [1]. AP includes acute edematous pancreatitis and acute necrotizing pancreatitis. Acute edematous pancreatitis has lower morbidity and hence, the clinical sequelae resolves without treatment. Whereas acute necrotizing pancreatitis has severe clinical sequelae, is more life threatening and has a higher rate of complications [2]. Notably, the incidence of AP is increasing, paralleled with an increase in the prevalence of obesity. Accumulation of visceral fat in patients with obesity increases the risk of severe AP through the action of adipocytokines, like, serum resistin. Recently, resistin has been described as serving as a marker of monocyte activation and it can be used to detect early inflammation of peri-pancreatic adipose tissue. Our study investigates the early rise of serum resistin and its correlation with the severity of acute pancreatitis before the full manifestation of the deleterious effects of acute pancreatitis.

## 2. Definition and Etiology of Acute Pancreatitis

Acute pancreatitis is a sudden inflammation of pancreas. Gallbladder stones are the most common cause of AP, representing close to 60% of all cases [3]. Several risk factors for AP have been proposed including alcohol, hyperlipidemia, hypercalcemia, drug consumption (furosemide, didanosine, estrogens, azathioprine, thiazide diuretics, sulfonamides, tetracyclines, sulindac, mercaptopurine, valproic acid, L-asparaginase) [4], Infections (mumps, mycoplasma, Epstein-Barr virus, *Ascaris lumbricoides*, HIV-related co infections), pancreatic cancer, pancreatic divism, trauma, autoimmune diseases (collagen vascular disease, immunoglobulin G4-related sclerosing acute pancreatitis), heredity, sphincter of oddi dysfunction, toxins, metabolic, post ERCP(2-3%) [5] and/or endocrine disorders.

A clear cause is not clinically determined in up to 30% of cases, which are referred to as idiopathic pancreatitis [6]. Acute pancreatitis (acute hemorrhagic pancreatic necrosis) is characterized by acute inflammation and necrosis of parenchyma, focal enzymic necrosis of pancreatic fat and vessel necrosis. This is caused by intrapancreatic activation of pancreatic enzymes. Lipase activation causes necrosis of fat tissue in pancreatic interstitial tissue and peripancreatic spaces as well as vessel damage. Necrotic fat cells appear as shadows, lacking the nucleus, pink contours of cells, finely granular cytoplasm, and you may find calcium precipitates. Digestion of vascular walls results in thrombosis and hemorrhage. Due to the pancreas lacking a capsule, the inflammation and necrosis can extend to include fascial layers in the immediate vicinity of the pancreas.

Several theories attempt to explain the pathogenesis of AP, whether it is a single or multifactorial pathologic entity. It has been proposed that gallstone migration into the common bile-pancreatic duct is the main cause of AP [7].

In the bulk of literature, the terms of biliary sludge and microlithiasis are often used interchangeably. Often, most patients with gallbladder stone induced biliary pancreatitis recover without significant sequelae. A large amount of visceral fat surrounding the pancreas may easily exacerbate the inflammation in AP and this hyperinflammatory state may enhance gallstone formation and also contribute to papillitis and sphincter dysfunction [8]. Smoking has also been reported to serve as a risk factor for AP [9]. Nicotin and its metabolites, especially cotinine have been directly implicated [10]. This substance is able to activate cell surface receptors on the exocrine pancreas, including nicotinic, acetylcholine receptor and mediate pancreatitis, therefore smoking is an established causal factor in inducing acute pancreatitis (AP) [11]. Injured endothelial cells may secrete vasoconstricting factors and reduce pancreatic microcirculation, which play a crucial role in pathogenesis of AP.

### 2.1. Ischemic Theory

It has been postulated that the excess fat in obese patients causes a decrease in the microcirculation. The incidence of acute pancreatitis (AP) is rising with increased prevalence of obesity, which exacerbates pancreatic injury. Microcirculatory disturbance play a pivotal role in pathogenesis. Hyperchylomicronemia causes sluggish flow in capillaries in the pancreatic bed and this increase ischemia. Ischemia then activates the inflammatory cascade and this worsens the cell damage.

*Pancreatic ischemia due to hypo perfusion may also be the critical factor which causes the progression from edema to necrosis [12].

## 3. Relation of Acute Pancreatitis with Obesity

Notably, the incidence of AP is increasing, paralleled with an increase in the prevalence of obesity, however, obesity may also be linked to the severity of AP. It is generally accepted that obesity is associated with an increased risk of AP development, therefore, larger abdominal adiposity and higher waist circumference (WC)often accompany severe pancreatitis. Obesity is a chronic low-grade inflammatory state characterized by high circulating levels of proinflammatory cytokines [13].

In fact, a new severity scoring system (acute physiology and chronic health evaluation-obesity) APACHE-O was proposed to consider obesity as an independent predictor for the outcome of AP [14].

Alternatively, obesity may intensify the immune response, which is able to exacerbate pancreatic injury. A study proposed by De Waele et al.[15]proposed that both peripancreatic and intrapancreatic fat and inflammatory cell can explain the high incidence of pancreatic inflammation and necrosis in patients with obesity. Furthermore, pancreatic microcirculation is lower in patients with obesity compared with those not obese, which increases the risk of ischemic injury, peripancreatic fat necrosis, and subsequent local infections. Obesity is associated with fatty infiltration in the pancreas and has been found to be inversely associated with the prognosis of AP[16]. It has been demonstrated that obesity is correlated with severity of AP, local complications, and mortality in patients with AP. Furthermore, patients who are overweight, or obese, are more vulnerable for initial acute attack of gallbladder stone induced pancreatitis [17]. The conclusion was that obesity is not only associated with an increased risk of AP development, but it is also a poor prognostic factor for AP [17].

Waist circumference was found to be an independent risk factor for the development of an initial attack of AP [18]. A large amount of visceral fat surrounding the pancreas may easily spread the inflammation in AP [19] and a strong correlation has been determined between peripancreatic visceral adipose tissue volume and AP severity.

## 4. Diagnostic Tools in Acute Pancreatitis

A routine medical history and clinical examination were conducted for each participant. The diagnosis of AP was established if 2 of the following 3 criteria were present:

1. Sudden onset of upper abdominal pain and/or abdominal localized signs characteristic of AP.
2. Findings on the abdominal computed tomography(CT) scan are consistent with AP.
3. Serum amylase and/or lipase elevation of at least 3 times the upper limit of normal level.

Gallstone pancreatitis was also diagnosed based on radiologic findings of a gallstone or bile duct dilatation and laboratory findings of obstructive jaundice. The presence of gallstones and dilatation of the extrahepatic biliary tree on cross-sectional abdominal imaging lend further support to the diagnosis of gallstone pancreatitis.

Several tests can help differentiate gallbladder stone pancreatitis from other causes of pancreatitis such as aspartate aminotransferase, alanine aminotransferase [20]. Alkaline phosphatase, and serum bilirubin which are the standard liver function tests including serum alanine transaminase (≥ 65 U/L), serum bilirubin level (> 18.1 Umol/L), and alkaline phosphatase (>137 U/L), should be reviewed before making a confident diagnosis. A combination of ultrasonography and blood tests yield a sensitivity of 95% to 98% [21]. Double-contrast material-enhanced CT scan after 48 hours post hospital admission was used to predict pancreatic parenchymal injury.

When patients have a severe contrast allergy or renal failure. Abdominal MRI without gadolinium can be used.

## 5. Scoring System of Acute Pancreatitis

Although Acute Physiology and Chronic Health Evaluation II (APACHE II)-[Table 1], Ranson [Table 2], Imrie, and Glasgow scoring system, along with Atlanta criteria [Table 3], were adopted to determine the severity of AP, this determination remains unpredictable. In fact, a new severity scoring system (acute physiology and chronic health evaluation-obesity) (APACH-O) was proposed to consider obesity as an independent predictor for the outcome of AP [14]. In addition, if the Ranson score was greater than 3, the APACHE II Score was 8 and above and the CT severity index was 5 and above, pancreatitis was considered severe [14]. An Atlanta criterion was adopted to define mild AP as (normal enhancement of pancreatic parenchyma on CT scan or if the Ranson score was less than 3), i.e., absence of the criteria of severe AP [22]. Whereas, severe AP was defined as the presence of at least one of the following criteria: pancreatic necrosis > 30%,Pancreatic abscess, pseudo cyst, systolic blood pressure < 90 mmHg, creatinine > 2 mg/dl after rehydration, gastrointestinal bleeding > 500 ml in 24 h, or death [22,23].

**Table 1. APACHE II Score (Acute Physiology and Chronic Health Evaluation II)**

| **APACHE II** | | | | | | | | | |
|---|---|---|---|---|---|---|---|---|---|
| **Punctuation APACHE II** | | | | | | | | | |
| **APS** | **4** | **3** | **2** | **1** | **0** | **1** | **2** | **3** | **4** |
| Tª rectal (°C) | ≥40.9 | 30–40.9 | | | 36–38.4 | | 32–33.9 | 30–31.9 | ≤30 |
| Mean arterial pressure | ≥159 | 130–159 | 110–129 | | 70–109 | | 50–69 | | ≤50 |
| Heart Rate | ≥179 | 140–179 | 110–129 | | 70–109 | | 55–69 | 40–54 | ≤40 |
| Breathing frequency | ≥49 | 35–49 | | 25–34 | 12–24 | 10–11 | 6–9 | | <6 |
| Oxygenation:<br>Si $FiO_2 \geq$ 0,5 ($AaDO^2$)<br>Si $FiO_2 <$ 0, 5 ($paO^2$) | ≥499 | | 200–349 | | <200<br><br>>70 | 61–70 | | 56–60 | <56 |
| pH arterial | ≥7.69 | | | | | | | | <7.15 |
| In Plasma (mmol/I) | ≥179 | 160–179 | 155–159 | 150–154 | 130–149 | | 120–129 | 111–119 | ≤111 |
| Plasma K (m mol/I) | ≥6.9 | 6.0–6.9 | | 5.5–5.9 | 3.5–5.4 | | 2.5–2.9 | | <2.5 |
| Creatinina * (mg/dI) | ≥3.4 | 2–3.4 | 1.5–1.9 | | 0.6–1.4 | | <0.6 | | |
| Hematocrit (%) | ≥59.9 | | 50–59.9 | 46–49.9 | 30–45.9 | | 20–29.9 | | <20 |
| Leukocytes (X 1000) | ≥39,9 | | 20–39,9 | 15–19,9 | 3–14,9 | | 1–2,9 | | <1 |
| Sum of points APS | | | | | | | | | |
| Total APS | | | | | | | | | |
| 15 – GCS | | | | | | | | | |

**Table 1. (Continued)**

| Age | Punctuation | Chronic disease | | | Points APS (A) | Points GCS (B) | Datum (C) | Previous disease points (D) |
|---|---|---|---|---|---|---|---|---|
| ≤44 | 0 | Post-operative program-ming | 2 | | | | | |
| 45–54 | 2 | Post-operative medical | 5 | | Total points APACHE II (A + B + C + D) | | | |
| 55–64 | 3 | ***Chronic disease***<br>*Hepatic*: cirrhosis (biopsy) or portal hypertension or previous episode of hepatic failure<br>*Cardiovascular*: dyspnea or resting angina (NYHA class IV)<br>*Respiratory*: Severe COPD, with hypercapnia, erythrocytosis or pulmonary arterial hypertension<br>*Renal*: chronic dialysis<br>*Immunocompromised*: treatment for immunosuppressive diseases | | | | | | |

* Creatinina: doblepuntuacónsi FRA

Source:(Viernes, April 3, 2009) http://pancreatititis.blogspot.hk.

**Table 2. Ranson's score**

| **Ranson's criteria** |
|---|
| At admission or diagnosis |
| Age over 55 years |
| White blood count over 16,000/cu mm |
| Blood glucose over 200 mg/dl |
| Serum Lactic dehydrogenase (LDH) over 350 U/l |
| Serum glutamic oxaloacetic transaminase (AST) over 250 U/l |
| During initial 48 h |
| Hematocrit fall greater than 10% points |
| Blood Urea nitrogen rise more than 5 mg/dl |
| Arterial $Po_2$ below 60 mm Hg |
| Serum calcium below 8 mg/dl |
| Base deficit > 4 meq/l |
| Estimated fluid sequestration more than 6000 ml |

Source: https://www.memorangapp.com/flashcards/79410/Criteria+and+Scores+for+ Pimps.

**Table 3. Atlanta criteria**

| **Atlanta Classification (Defining Severe Acute Pancreatitis)** |
|---|
| • **Evidence of Organ Failure** |
| Shock (Systolic Blood Pressure < 90 mm Hg) |
| Pulmonary insufficiency (PaO2 < 60 mm Hg) |
| Renal failure (creatinine > 500 ml/day) |
| • **Or Local Complications** |
| Pancreatic necrosis > 30% |
| Pancreatic abscess |
| Pancreatic pseudocyst |
| • **With unfavourable Prognostic Signs** |
| Ranson Criteria ≥ 3 or |
| APACHE II score ≥ 8 |

Source: (http://www.testlllnutrition.com/mod_lll/TOPIC14/m141new.htm).

Computed tomography contrast-enhanced abdominal scans were used to predict pancreatic parenchymal injury 48 hours after hospital admission.

We graded the severity of pancreatitis radiologically into five distinct groups from A to E according to the Balthazar-Ranson scoring system[24].

## 5.1. CT Calculated Severity Index

### *5.1.1. CT Acute Pancreatitis Grading*

| | | |
|---|---|---|
| A. | Normal pancreas | 0 |
| B. | Enlarged pancreas | 1 |
| C. | Inflamed pancreas with or peri-pancreatic fat inflammation | 2 |
| D. | Single peri-pancreatic fluid collection | 3 |
| E. | Two or more peri-pancreatic collections | 4 |

### *5.1.2. CT - Grading of Acute Pancreatitis Necrosis*

#### 5.1.2.1. Necrosis percentage additional points

| | |
|---|---|
| 0% | 0 |
| 30-50% | 4 |
| >50% | 6 |

**CT severity index is calculated by adding CT grade points to points assigned for percentage of pancreatic necrosis (from 0 to 10.)*

***CT SEVERITY INDEX = ___ +___ =___(from 0-10)***

clinical severity evaluation was carried out using the Ranson and APACHE scores within 48 h after admission.

Furthermore, late complications like pancreatic pseudocyst and abscesses were followed up in the outpatient clinic after 4-8 weeks by computed tomography abdominal scan based on clinical presentation.

## 6. New Predictive Inflammatory Markers in Acute Pancreatitis

Experimental biochemical markers have been studied to assess disease severity, including trypsinogen an activation peptide, interleukin-6, interleukin-10, procalcitonin, phospholipase A2 and C-reactive protein [25].C-reactive protein (CRP) is a commonly used marker for distinguishing between mild and severe acute pancreatitis attack. Damage to the pancreas in acute pancreatitis and the intensity of the organism response (i.e., the acute-phase response) is accompanied by a substantial increase in the serum CRP level, which is the most significant reactant in this response, as a result of hepatocyte stimulation by cytokines. In acute inflammation, CRP levels often reach their peak at 48 hours [26].

It was concluded that obesity increases the severity of AP by amplifying the immune response to injury. The increased amount of adipose tissue in these individuals promotes a low-grade, systemic inflammatory state [27]. In line with this data, obesity is also associated with higher levels of inflammatory markers.

In AP, pancreatic injury leads to a massive release of pancreatic lipase that causes digestion of peri-pancreatic adipose tissue which becomes infiltrated by significant quantities of monocytes. This process leads to the synthesis and secretion of adipose-specific proteins, termed adipocytokines [28] which serve as new predictive markers [29], with potent immunomodulatory and metabolic activities. The metabolic and pro-inflammatory changes seen in AP might be partially caused by these proteins such as leptin, adiponectin, Ghrelin, apelin and resistin.

Changes in the function and quantity of these proteins play a role in the pathogenesis and progress of inflammation, inflammatory response, insulin resistance and metabolic syndrome.

One of the novel adipocytokines, resistin, is produced by white and brown adipose tissues [30]. Resistin is a newly identified peptide hormone, secreted specifically [31, 32] by adipocytes, that can cause obesity and hypertriglyceridaemia, due to its association with insulin resistance. Studies have revealed that resistin is also an important cytokine in inflammatory reactions, and in the regulation of other cytokines. Resistin has been identified in several other tissues, including the islets of pancreas[33], hypothalamus, pituitary and adrenal glands, gastrointestinal tract, myocytes, spleen, white blood cells and plasma. Resistin antagonizes insulin action, and is downregulated by rosiglitazone and peroxisome proliferator-activated receptor gamma agonists [34].Resistin can act as prooxidant mediator for endothelial dysfunction by release of Endothelin-1 as a marker of endothelial dysfunction [35].

Recently, resistin has been described as serving as a marker of monocyte activation [36],and it can be used to detect early inflammation of peri-pancreatic adipose tissue. Our study investigates the ability of serum resistin levels upon hospitalization to predict clinical severity of AP.

Importantly, human resistin mRNA expression is higher in abdominal adipose tissue than in subcutaneous adipose tissue [37], therefore we chose to investigate the serum level of resistin in AP patients to assess its potential value in detecting pancreatitis severity specifically induced by adipose tissue inflammation. Until now, resistin has not been reported in purely acute biliary pancreatitis patients of a substantial sample size.

## 7. Methodology of Using Resistin

A group of 112 patients with an initial attack of acute biliary pancreatitis were included in this prospective study between January 2011 and Feb 2014. They were admitted to the surgical ward at Ohod Hospital, Saudi Arabia. Study protocols were approved by an institutional review board. All patients provided written informed consent and underwent a routine medical history and clinical examination. The diagnosis of AP was based on acute upper abdominal pain with an elevated serum amylase of at least three times the upper normal limit, which peaks approximately 24 hours after the onset of attack and/or serum lipase which is more sensitive and specific in the diagnosis of pancreatitis.

None of the patients had ever consumed alcohol and none had history of drug consumption known to cause pancreatitis.

Serum calcium and lipid profiles were tested to exclude hypercalcemia or hyperlipidemia-induced pancreatitis. Patients without obvious cause of AP underwent endoscopic retrograde cholangiopancreatography (ERCP)

## 8. Exclusion Criteria

Patients were excluded from the study based on the following;

1. Diabetes mellitus
2. Presence of morbidities possibly affecting patient weight, such as hypothyroidism or pregnancy
3. Alcohol consumption
4. History of episodes of idiopathic and chronic recurrent pancreatitis.
5. Drug induced pancreatitis
6. Pancreatic divism (is a congenital anomaly in the anatomy of the ducts of the pancreas in which a single pancreatic duct is not formed, but rather remains as two distinct dorsal and ventral ducts).
7. ERCP induced pancreatitis
8. Hyperlipidemia and hypercalcemia induced pancreatitis
9. Patients with symptoms for more than 48 hours..

Venous blood samples were collected from the antecubital vein of patients into vacuum tubes containing EDTA or a serum separator gel. After sampling, the tubes were immediately centrifuged at 1.5gfor 10 minutes. Aliquots of serum were stored at - 20°C, and routine hematology and biochemistry tests were performed 48 hours after hospital admission while the patient was fasting. All biochemical measures were carried out by the same team of laboratory technicians using the same methods throughout the study period.

Samples were collected from AP patients at time of hospital admission. We enrolled a control group of 112 persons with matched BMI, gender and age, because serum resistin concentration increases in obesity, and is positively correlated with BMI, insulin resistance or body fat[38].

Resistin plasma samples for both test patients and control groups were stored at -80°C and diluted 100-fold.until biochemical analysis was completed

One technician performed the analysis in a blinded fashion. Resistin concentrations for both groups were measured by using ELISA.

Anthropometric measurement were determined using BMI and WC. Obesity was classified by BMI[Table 4], which was measured by weight in kilograms divided by height in meters squared (kg/m$^2$)

As recommended by the World Health Organization[39],we established the BMI cutoff points to define:

**Table 4. BMI classification**

| Weight | BMI |
|---|---|
| Normal body weight | 18.5-24.9 kg/m2 |
| Overweight | 25-29.9 kg/m2 |
| Obesity class1 | 30-34.9 kg/m2 |
| Obesity class 2 | 35.0-39.9 kg/m2 |
| Obesity class 3(Morbid) | 40 kg/m2≥ |

Source: https://www.circlesurrogacy.com/blog/circle-surrogacy/the-importance-of-body-mass-index/

Waist circumference was measured in the standing position at the height of the umbilicus while the subject breathed out gently. When marked fat accumulation caused the umbilicus to sag downward, measurements were taken at a height midway between the lower margin of the ribs and the anterior iliac spine using a circumference measuring tape (Seca 200, Hamburg, Germany).

## 9. ELISA Assay

Serum was quantitatively assayed for resistin using a sandwich enzyme immunoassay.

Standards and samples were sandwiched by immobilized antibody and biotinylated polyclonal antibodies specific for resistin. Samples were analyzed using the Bio-plex suspension array system, which included a fluorescent reader and Bio-plex Manager analytic software at the Taibah University Biochemical Department. Mean values of duplicate readings were calculated for each standard and sample in a blinded fashion. The cut-off value of resistin was 15 ng/mL.

## 10. Results

Our study consisted of 112 patients (67 females and 45 males) diagnosed with an initial first attack of acute gallbladder-stone induced pancreatitis. The mean age was 46 years/and mean BMI value was 31.4 kg/m$^2$ (Obese, class 1).

During the first 48 hours after admission 27 patients (24.1%) suffered from severe AP, but the majority 85 patients experienced mild pancreatitis 75.9%, all of whom were diagnosed based on the Atlanta criteria.

- *Local complications* (pancreatic pseudocyst, pancreatic necrosis, acute fluid collection, pancreatic abscess and thrombosis), while
- *systemic complications*; shock, pulmonary insufficiency, renal failure gastrointestinal bleeding, disseminated intravascular coagulation (DIC) and severe metabolic disturbances occurred in 26 (23.2%) and 22 (19.6%) patients, respectively. The observed mortality rate was 4.4% and we classified the complications as either early or late. Early complications (shock, pulmonary insufficiency, renal failure, gastrointestinal bleeding, DIC, severe metabolic disturbance and acute fluid collection) manifest at the onset or within the first 2–3 days, whereas, late complications (pancreatic necrosis, pseudocyst, pancreatic abscess and thrombosis) usually manifest months or years following the resolution of an acute attack.

**Table 5. Patient Characteristics (N=112)**

| Patient Characteristics | Result |
|---|---|
| **Age - years**<br>(Mean ± SD) | 46 ± 16.5 |
| **Severity of Pancreatitis**<br>(a) - Mild<br>(b) - Severe | (a) - (85)75.9%<br>(b) - (27)24.1% |
| **Gender** (male/female) | (45/67) 40.1%/59.8% |
| **Central Obesity**<br>**Waist Circumference - cm**<br>(Mean ± SD) | 95.2± 14.65 |
| **Body Mass Index - Kg/m²**<br>(Mean ± SD)<br>(a) – Underweight%<br>(b) – Normal%<br>(c) – Overweight%<br>(d) – Obese%<br>(e) - Morbid obesity% | 31.4 ± 5.75<br>(a) - (3) 2.6%<br>(b)- (22)19.6%<br>(c) - (25)22.3%<br>(d) - (49)43.7%<br>(e) - (13)11.6% |
| **Complications**<br>**(a) – Local**<br>1. Acute fluid collection<br>2. Pancreatic necrosis<br>3. Pancreatic pseudocyst<br>4. Pancreatic abscess<br>5. Thrombosis<br>**(b)- Systemic**<br>1. Shock<br>2. Pulmonary insufficiency<br>3. Renal failure<br>4. GIT bleeding<br>5. DIC<br>6. Severe metabolic disturbance | <br>(a) - (26)23.2%<br>1. (12)46.1%<br>2. (9)34.6%<br>3. (5)19.2%<br>4. (0)0%<br>5. (0)0%<br>(b)- (22)19.6%<br>1. (4)18.1%<br>2. (10)45.4%<br>3. (2)9.1%<br>4. (2)9.1%<br>5. (2)9.1%<br>6. (2) 9.1% |
| **Death** (Mortality) | (5)4.4% |

Our results showed that only BMI significantly correlated with severity of acute gall stone pancreatitis (P = 0.007; Table 6), (Figure 1).

Serum resistin concentration was significantly higher in patients with AP compared with control subjects (Figure 2). Resistin concentrations upon hospital admission correlated with AP. Receiver-operator curve and area under the curve (AUC) delineate sensitivity and specificity of serum resistin concentration as an inflammatory marker on the day of hospital admission in AP, AUC>97% and cutoff value: 15 ng/mL (Figures3A and B).

**Table 6. Relationship between severity of pancreatitis and BMI (N=112)**

| Patient characteristics | Total number of patients (N=112) (100%) | Mild Pancreatitis (N=85) 75.9% | Severe Pancreatitis (N=27) 24.1% | Chi-square |
|---|---|---|---|---|
| **Body Mass Index(Kg/m²)** | | | | |
| A -Underweight | 3(2.6%) | 3(3.5%) | 0 (0%) | **0.007** |
| B -Normal | 22 (19.6%) | 19(22.3%) | 3(11.1%) | |
| C -Overweight | 25 (22.3%) | 15(17.6%) | 10(37%) | |
| D –Obese | 49 (43.7%) | 42(49.4%) | 7(25.9%) | |
| E –Morbid obesity | 13 (11.6%) | 6(7%) | 7(25.9%) | |

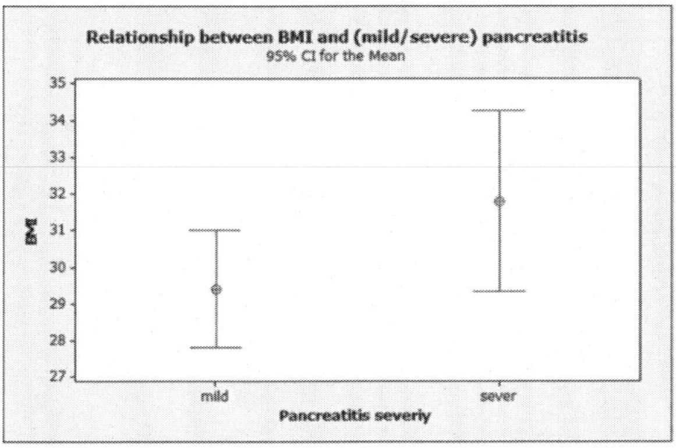


Figure 1. Relationship between obesity and pancreatitis severity.

We investigated the local and systemic complications and their relationship with BMI and reported that acute fluid collection and pancreatic pseudocysts correlated significantly with obesity (P=0.043 and P=0.000) respectively.

In addition, a CT severity index of ≥5 was noticeable in obese Patients (P=0.032).

Central fat distribution was revealed as a more definitive risk factor than the actual amount of fat and it is, therefore plausible that the inflammatory properties of intrapancreatic fat may be involved in triggering the onset of AP.

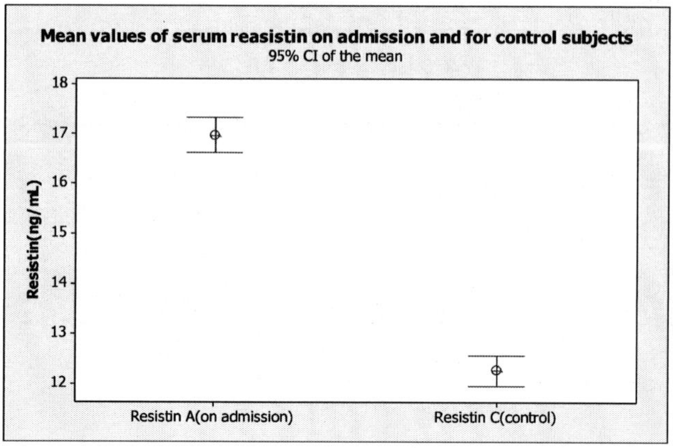


Figure 2. Significant difference between resistin level in acute pancreatitis patients and control subjects (P = 0.000).

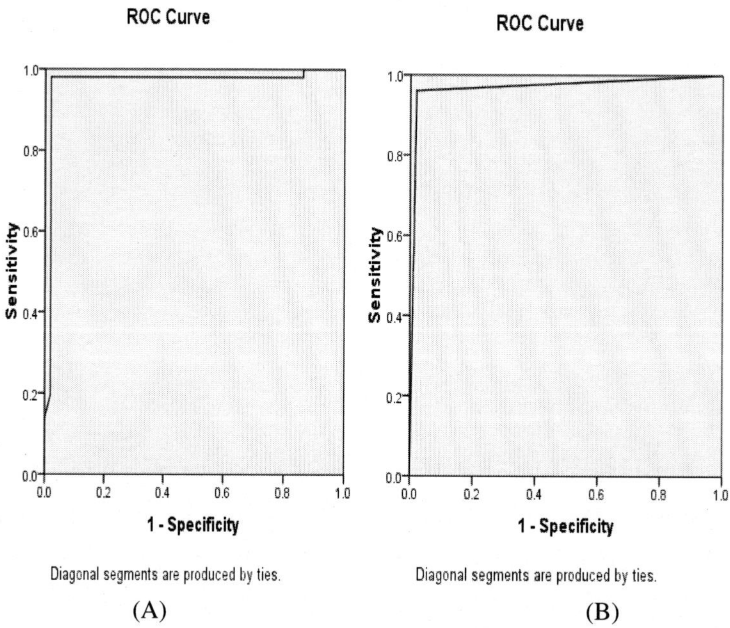


Figure 3. (A, B). Receiver-operator curve and area under the curve. A: Receiver-operator curve (ROC) and area under the curve (AUC) delineate sensitivity and specificity of serum resistin concentration (day of admission), AUC > 97%, *P*-value < 0.0001; B: ROC and AUC of serum resistin value on admission. Cutoff value: 15 ng/mL and positive predictive value: 97%.

Patients with an android fat distribution and higher Waist circumference measurements are at greater risk of developing severe AP[40].

De Waele et al. [15] demonstrated that obesity is correlated with severity of AP, local complications and mortality in patients with AP and these results were confirmed via meta-analysis carried out by Wang et al.[41].Furthermore, our study confirmed that a BMI of >31kg/m² is associated with severe pancreatitis(P=0.007) (see Table 6).

## 11. Clinical Application

A majority of patients in our study had acute gallbladder – stone induced pancreatitis. Our results showed that:

1. Obesity significantly correlated with severity of acute biliary pancreatitis.
2. Serum resistin concentration was significantly higher in patients with AP compared with control subjects.

Exploration of an inflammatory marker such as resistin, reflects the severity burden of AP and is superior over other scoring approaches.

## 12. Other Studies

Resistin levels are increased in the pancreatic tissues of patients with AP, and the increased expression has been shown to correlate with the severity of AP [42, 43].Moreover, LE-NING XUE et al. [44], the mRNA and protein expression levels of resistin increased more notably in the acute necrotizing pancreatitis group compared with the acute edematous pancreatitis group, indicating that resistin is activated in the pathogenesis of AP and is involved in pancreatic tissue damage.

Leniowski et al. [45] measured serum resistin in 30 AP patients and concluded that the serum concentration of resistin may represent a useful additional marker of inflammatory response in AP.

Serum levels of resistin have been reported in 50 patients with AP, showing that 40%had biliary pancreatitis, the level of resistin correlated with Ranson and APACHE scores[36], suggesting that resistin concentration upon hospital admission may serve as an early predictive marker of peri-pancreatic necrosis and clinical severity for AP.

However, these studies used a small sample size and 40% of the patients had biliary pancreatitis with resistin levels (measured over the first 10 days after admission-mean value),which is not reflective of the single initial value alone.

Our study, however, had twice the sample size and most of the pancreatitis cases were biliary in origin, which has not previously been extensively studied. Additionally, we measured the serum resistin levels in our patients within a few hours after a confirmed diagnosis of acute biliary pancreatitis upon hospital admission.

Our study was designed with strict exclusion criteria to eliminate any confounding variables that affected patients weight or BMI. Also, we used a biochemical marker that specifically reflect the inflammation of peri-pancreatic adipose tissue. We noticed that a large fraction of our patients had acute biliary pancreatitis and did not consume alcohol, leaving the possibility that alcoholic AP might behave quite differently. Adrych et al. [46] demonstrated that serum resistin concentration is elevated in a small cohort of patients with chronic pancreatitis of alcoholic origin (n=50).

## 13. Recommendations

Before resistin can be used widely in the clinical setting, our results must be confirmed by larger and multi-center studies.

The use of new pancreatic injury markers of inflammatory response appears very promising, although most have been studied mainly in a research setting so their clinical impact remains to be determined.

Resistin may represent a new, inexpensive and effective indicator for predicting the severity of acute pancreatitis and presence of necrosis.

## CONCLUSION

In conclusion, resistin was demonstrated to be activated in the pathogenesis of AP and involved in the injury process of pancreatic tissue. The expression of resistin was found to be associated with the severity of the damage in the pancreatic tissue. Therefore, resistin may be used as an indicator to assess the severity and prognosis of AP, and the hormone may prove a prime target for the future therapeutic interventions.

Resistin serum levels can be used as a diagnostic marker for AP, as levels are increased upon hospital admission.

## REFERENCES

[1] Abu-Zidan, FM; Bonham, MJ; Windsor, JA. Severity of acute pancreatitis: a multivariate analysis of oxidative stress markers and modified Glasgow criteria. *Br J Surg.*, 2000, 87, 1019-1023.

[2] Wu, BU; Banks, PA. Clinical management of patients with acute pancreatitis. *Gastroenterology.*,2013, 144, 1272-1281.

[3] Andersson, R; Andersson, B; Haraldsen, P; Drewsen, G; Eckerwall, G. Incidence, management and recurrence rate of acute pancreatitis. *Scand J Gastroenterol.*, 2004, 39, 891-894.

[4] Brunicardi, FC; Andersen, DK; Billiar, TR. Chapter 32: Pancreas. In: *Schwartz's principles of surgery*. 8th ed. New York, NY: McGraw-Hill; 2005.

[5] Cheng, CL; Sherman, S; Watkins, JL; Barnett, J; Freeman, M; Geenen, J; Ryan, M; Parker, H; et al. Risk factors for post-ERCP pancreatitis: a prospective multicenter study. *Am J Gastroenterol.*, 2006, 101, 139-47.

[6] Ballinger, AB; Barnes, E; Alstead, EM; et al. Is intervention necessary after a first episode of acute idiopathic pancreatitis? *Gut.*, 1996, 38, 293-295.

[7] Steer, LM. Etiology and pathogenesis of acute pancreatitis. *Ann Ital Chir.*, 1995, 66, 159-163.

[8] Abeysuriya, V; Deen, KI; Navarathne, NM. Biliary microlithiasis, sludge, crystals, microcrystallization, and usefulness of assessment of nucleation time. *Hepatobiliary Pancreat Dis Int*, 2010, 9, 248–53.

[9] Morton, C; Klatsky, AL; Udaltsova, N. Smoking, coffee, and pancreatitis. *Am J Gastroenterol.*, 2004, 99, 731-738.

[10] Edderkaoui, M[1]; Thrower, E. "Smoking and pancreatic disease," *J Cancer Ther.*, 2013, 4, 34-40.

[11] Barreto, SG. "How does cigarette smoking cause acute pancreatitis?," *Pancreatology*, 2016, 16, 157-63.

[12] Warshaw, AL; O'Hara, PJ. Susceptibility of the pancreas to ischemic injury in shock. *Ann Surg.*, 1978, 188, 197-201.

[13] Lee, YH; Pratley, RE. The evolving role of inflammation in obesity and the metabolic syndrome. *Curr Diab Rep.*, 2005, 5, 70-75.

[14] Papachristou, GI; Papachristou, DJ; Avula, H; et al. Obesity increases the severity of acute pancreatitis: performance of APACHE-O score and correlation with the inflammatory response. *Pancreatology.*, 2006, 6, 279-285.

[15] De Waele, B; Vanmierlo, B; Van Nieuwenhove, Y; et al. Impact of body overweight and class I, II and III obesity on the outcome of acute biliary pancreatitis. *Pancreas.*, 2006, 32, 343-345.

[16] Johnson, CD; Toh, SK; Campbell, MJ. Combination of APACHE-II score and an obesity score (APACHE-O) for the prediction of severe acute pancreatitis. *Pancreatology.*, 2004, 4, 1-6.

[17] Sawalhi, S; Al-Maramhy, H; Abdelrahman, AI; Allah, SE; Al-Jubori, S. Does the presence of obesity and/or metabolic syndrome affect the course of acute pancreatitis?: A prospective study. *Pancreas*, 2014, 43, 565-70.

[18] Sadr-Azodi, O; Orsini, N; Andre´n-Sandberg, A; et al. Abdominal and totaladiposity and the risk of acute pancreatitis: a population-

basedprospective cohort study. *Am J Gastroenterol.*, 2013, 108, 133-139.

[19] Yashima, Y; Isayama, H; Tsujino, T; et al. A large volume of visceraladipose tissue leads to severe acute pancreatitis. *J Gastroenterol.*, 2011, 46, 1213-1218.

[20] Wang, SS; Lin, XZ; Tsai, YT; et al. Clinical significance of ultrasonography, computed tomography, and biochemical tests in the rapid diagnosis of gallstone-related pancreatitis: a prospective study. *Pancreas.*, 1988, 3, 153-158.

[21] Ammori, BJ; Boreham, B; Lewis, P; Roberts, SA. The bio chemical detection of biliary etiology of acute pancreatitison admission: a revisit in the modern era of biliary imaging. *Pancreas*, 2003, 26, e32-e35.

[22] Bollen, TL; van Santvoort, HC; Besselink, MG; et al. The Atlanta classification of acute pancreatitis revisited. *Br J Surg.*, 2008, 95, 6-21.

[23] de-Madaria, E; Soler-Sala, G; Lopez-Font, I; Zapater, P; Martínez, J; Gómez-Escolar, L; Sánchez-Fortún, C; Sempere, L; Pérez-López, J; Luís, F; Pérez-Mateo, M. Update of the Atlanta Classification of severity of acute pancreatitis: should a moderate category be included? *Pancreatology*, 2010, 10, 613-9.

[24] Balthazar, EJ; Ranson, JH; Naidich, DP; Megibow, AJ; Caccavale, R; Cooper, MM. Acute pancreatitis: prognostic value of CT. *Radiology*, 1985, 156, 767-772.

[25] Smotkin, J; Tenner, S. Laboratory diagnostic tests in acute pancreatitis. *J Clin Gastroenterol*, 2002, 34, 459-462.

[26] Viedma, JA; Perez-Mateo, M; Dominguez, JE; et al. Role of interleukin-6 in acute pancreatitis. Comparison with C-reactive protein and phospholipase A. *Gut*, 1992, 33, 1264–1267.

[27] Sempere, L; Martinez, J; de Madaria, E; et al. Obesity and fat distribution imply a greater systemic inflammatory response and a worse prognosis in acute pancreatitis. *Pancreatology.*, 2008, 8, 257-264.

[28] Schäffler, A; Müller-Ladner, U; Schölmerich, J; Büchler, C. Role of adipose tissue as an inflammatory organ in human diseases. *Endocr Rev*, 2006, 27, 449-467.

[29] Schäffler, A; Schölmerich, J; Salzberger, B. Adipose tissue an immunological organ: Toll-like receptors, C1q/TNFs and CTRPs. *Trends Immunol*, 2007, 28, 393-399.

[30] Steppan, CM; Lazar, MA. The current biology of resistin. *J Intern Med*, 2004, 255, 439-447.

[31] Beltowski, J. Adiponectin and resistin–new hormones of white adipose tissue. *Med Sci Monit*, 2003, 9, RA55–RA61.

[32] Fujita, H; Fujishima, H; Morii, T; Koshimura, J; Narita, T; Kakei, M; Ito, S. Effect of metformin on adipose tissue resistin expression in db/db mice. *Biochem Biophys Res Commun.*, 2002,298, 345-9.

[33] Minn, AH; Patterson, NB; Pack, S; Hoffmann, SC; Gavrilova, O; Vinson, C; Harlan, DM; Shalev, A. Resistin is expressed in pancreatic islets. *Biochem Biophys Res Commun*, 2003, 310-641.

[34] Adeghate, E. An update on the biology and physiology of resistin. *Cell Mol Life Sci.*, 2004, 61(19-20), 2485-96.

[35] Ściskalska, M; Marek, G; Grzebieniak, Z; Milnerowicz, H. Resistin as a Prooxidant Factor and Predictor of Endothelium damage in Patients with Mild Acute Pancreatitis Exposed to Tobacco Smoke Xenobiotics. *Mediators Inflamm.*, 2017, 2017, 3039765.

[36] Schäffler, A; Hamer, O; Dickopf, J; Goetz, A; Landfried, K; Voelk, M; Herfarth, H; Kopp, A; Büchler, C; Schölmerich, J; Brünnler, T. Admission resistin levels predict peripancreatic necrosis and clinical severity in acute pancreatitis. *Am J Gastroenterol*, 2010, 105, 2474-2484.

[37] McTernan, PG; McTernan, CL; Chetty, R; Jenner, K; Fisher, FM; Lauer, MN; Crocker, J; Barnett, AH; Kumar, S. Increased resistin gene and protein expression in human abdominal adipose tissue. *J Clin Endocrinol Metab*, 2002, 87.

[38] Degawa-Yamauchi, M; Bovenkerk, JE; Juliar, BE; Watson, W; Kerr, K; Jones, R; Zhu, Q; Considine, RV. Serum resistin (FIZZ3) protein is increased in obese humans. *J Clin Endocrinol Metab*, 2003, 88, 5452-5455.

[39] Seidell, JC;Flegal, KM. Assessing obesity: classification and epidemiology. *Br Med Bull.*, 1997, 53, 238-252.

[40] Mery, CM; Rubio, V; Duarte-Rojo, A; et al. Android fat distribution as predictor of severity in acute pancreatitis. *Pancreatology*, 2002, 2, 543–9.

[41] Wang, SQ; Li, SJ; Feng, QX; Feng, XY; et al. Overweight is an additional prognostic factor in acute pancreatitis: a meta-analysis. *Pancreatology*, 2011, 11, 92–8.

[42] Daniel, P; Leniowski, B; Mokrowiecka, A; Jasiska, A; Pietruczuk, M; Małecka-Panas, E. Circulating levels of visfatin, resistin and proinflammatory cytokine interleukin.8 in acute pancreatitis. *Pancreatology*, 10, 477.482, 2010.

[43] Schäffler, A; Landfried, K; Volk, M;Fürst, A; Büchler, C; Schölmerich, J; Herfarth, H. Potential of adipocytokines in predicting peripancreatic necrosis and severity in acute pancreatitis: pilot study. *J Gastroenterol Hepatol*, 22, 326.334,2007.

[44] Xue, LN; Wang, XY; Tan, Y; Lin, M; Zhang, W; Xu, KQ. Significance of resistin expression in acute pancreatitis. *Exp Ther Med.*, 2015, 9, 1438-1442.

[45] Leśniowski, B; Kumor, A; Jasińska, A; Daniel, P; Pietruczuk, M; Małecka-Panas, E. [Resistin-- a new laboratory marker useful in diagnosis of acute pancreatitis?]. *Pol Merkur Lekarski*, 2007,22, 385-7.

[46] Adrych, K; Smoczynski, M; Sledzinski, T; Dettlaff-Pokora, A; Goyke, E; Swierczynski, J. Increased serum resistin concentration in patients with chronic pancreatitis: possible cause of pancreatic fibrosis. *J Clin Gastroenterol*, 2009, 43, 63.

## BIOGRAPHICAL SKETCH

***Samer AL Sawalhi***

**Affiliation:** Taibah University/Saudi Arabia, Dalhousie University/ Canada, Tongji University/China

**Education:**

(1997-2003) - Medical and surgical bachelor degree from IRAQ-Baghdad University.
(2004-2009)- Training-residency program in general and laparoscopic surgery-Jordan.
(2008-2010)- Membership in the Royal college of surgery (Ireland-Dublin).
(2008-2010)- Surgical Oncology training –King Hussein Cancer Center-Jordan.
(2011-2013)- Arab board in General and Laparoscopic surgery
(2013-2014)- Minimal Invasive Thoracic surgery Fellowship-Dalhousie University-Canada.
(2018-now)- Minimal Invasive Uniportal Thoracic surgery Fellowship-Tongji University-China

**Research and Professional Experience:** I worked as assistant professor and surgical specialist at Taibah University –Department of surgery, during which I focused my research on acute pancreatitis and I used for the first time a new inflammatory marker in my research (Resistin in acute pancreatitis). Furthermore, I expanded my research on this topic and I studied the relationship between Metabolic syndrome and incidence of acute pancreatitis. Moreover I conducted and shared in the establishment of the [Center of Genetics and Inherited Diseases].Taibah University-Al-Madina-KSA(2012)and I worked as a permanent researcher in that center (Inherited surgical oncology field)

**Innovation procedure:** We created a new technique in treating trachea-esophageal fistula, and this technique was accepted all over the world and it was distributed as a video in the *Journal of Thoracic and Cardiovascular surgery in the United states*- April-2007.

*Closure of tracheogastric fistula by video-assisted tracheoscopy, direct repair and self-expandable titanium stent in a patient with total laryngo pharyngo esophagectomy) which published in *Thoracic and Cardiovascular surgery journal*, Volume 133, Issue 4, April 2007, pages 1103-1104-Video On line

**Research Work:**

1-(Lambert Academic Publishing Company-Germany) considered publishing my research work about (Breast phyllodes tumor) as a printed book (ISBN: 978-3-659-42077-1) 3-March-2015: https://www.morebooks.de/store/gb/book/breast-phyllodes-tumor/isbn/978-3-659-42077-1

**Professional Appointments:**

- Surgical Oncology specialist –King Hussein Cancer Center
- Assistant Professor in Surgical Department-Taibah University
- Permanent researcher in the Center of Genetics and Inherited Diseases
- Taibah University-Saudia Arabia.
- Thoracic surgery Fellow-Dalhousie University-Canada
- Membership in European Society of Thoracic surgeon

**Honors:**

March 2011 - Award for conducting and sharing in the establishment of the [Center of Genetics and Inherited Diseases]. Taibah University-Al-Madina-Saudi Arabia

June 2011 – Award for best teacher in the college of medicine-Taibah University for the 4th year students
Taibah University-Al-Madina-Saudi Arabia

March 2013 - Award for excellence in scientific research-
Scientific Research Deanship-
Taibah University, Saudi Arabia
January 2014 - Award from Pan Arab health (Young surgeon of the year 2014)
Dubai-United Arab Emirates

**Publications from the Last 3 Years:**

[1] Sawalhi, Samer; Hamdi, Al-Maramhy; Salman, Al-Jubori; Abdelrahman, l-Abdelrahman; Salah Eldin; Geib Allah; et al. Role of Adiponectin in Gallbladder- a prospective case–control study. *Hamdan Medical Journal*, Vol 7, No 1 (2014).
[2] Sawalhi, Samer; Hamdi, Al-Maramhy; Abdelrahman, I Abdelrahman. Resistin is not good biochemical marker in acute biliary pancreatitis severity: prospective case – controlled study- *World journal of gastroenterology*, Nov-2014-15351-57.
[3] Sawalhi, Samer; Al-Maramhy, H; Abdelrahman, AI; Allah, SE; Al-Jubori, S; et al. Does the presence of obesity and/or Metabolic syndrome impact the course of acute pancreatitis? *Prospective study Pancreas Journal*. 2014 May, 43(4), 565-70.
[4] Samer, Al Sawalhi, Is it Feasible to Concomitantly Resect Gastric Gastrointestinal Stromal Tumor during Bariatric Laparoscopic Sleeve Gastrectomy? Case report. *Annals of clinical case report Journal*, 2017 | Volume 2 | Article 1346.
[5] Sawalhi, Samer; Al-Shatti, M. Phyllodes tumour of the breast: The impact of histopathological factors in local recurrence and distant metastasis. *Annals of Saudi Medicine journal*. 2013. Mar-Apr, 33(2), 162-8
[6] Sawalhi, Samer; Al-Harbi, K; Raghib, Z; Abdelrahman, AI; Al-Hujaily, A, et al. Behavior of advanced gastrointestinal stromal tumor

in a patient with Von-Recklinghausen disease: Case Report. *World journal of clinical oncology*, August-2013,10, 4(3), 70-4.

[7] Uniportal subxiphoid video-assisted thoracoscopic surgery, a feasible approach for a challenging middle lobectomy in an obese patient was accepted *on Journal Of visualized surgery*, July-2018.

# INDEX

## A

## B



## C

## D



## E

## F



## G



## H

## I



## K



## L

## P



## R



## S

## T

## U



## V



## W